Prentice Hall International Inc., London
Prentice Hall of Australia, Pty. Ltd., Sydney
Prentice Hall of Canada Ltd. Toronto
Prentice Hall of India Private, Ltd., New Delhi
Prentice Hall of Japan Inc., Tokyo
Prentice Hall Inc., Englewood Cliffs, New Jersey

PHYSICAL GEOGRAPHY
A Systems Approach

Richard J. Chorley

UNIVERSITY OF CAMBRIDGE

Barbara A. Kennedy

UNIVERSITY OF CAMBRIDGE

PHYSICAL GEOGRAPHY

A Systems Approach

Prentice-Hall International Inc., London

ISBN 13–669028–9 (paper)
     13–669036–x (cloth)

Current printing (last digit)
10 9 8 7 6 5 4 3 2 1

Printed at The Pitman Press, Bath, England

Preface

> The great forward thrusts of intellectual
> ordering and of practical manipulation have
> come from the creative rethinking of the basic
> entities to which we give our attention. In
> science we commonly think that quantitative
> precision is our great strength; and, certainly,
> it does give science its essentially unique power
> of precise prediction and hence control. But
> before measurements can be meaningful they must
> be directed to the right things and, even in
> science, finding these things is the major
> achievement; entitation is more important than
> quantification.
>
> R. W. GERARD: *Science*, 1965, p. 762.

This book has not been prepared as a text for any existing course in physical geography, although it may become one. Rather, it represents an attempt to crystallize an attitude to that part of reality with which physical geographers have been involving themselves. This attitude has developed over a considerable period, but recently the rate of change has been increasing and a number of previously separate themes are beginning to be unified. For example, one of the most important of these themes which is encouraging present changes emerged more than two decades ago, when Arthur N. Strahler involved his work in fluvial geomorphology with the general systems approach of Ludwig von Bertalanffy. The present book attempts to draw together many such strands and represents a synthesis of a wide variety of modern scholarship in many fields. For the sake of continuity in presentation, acknowledgement to these sources has been reserved to the references which are given at the end of each chapter.

In two important respects this book can make some claim to originality when it is compared with existing works on 'physical geography'. Firstly, it represents an unreserved attempt to show how the phenomena of physical geography can be rationalized and perhaps made to assume new significance and coherence when treated in terms of systems theory, statistical analysis, cybernetics, and other modern inter-disciplinary approaches to the features of the real world. To assist in this respect, an extensive glossary of terms is provided. Secondly, the book involves itself with the increasing problem of the contemporary definition of physical geography, although this is done by example rather than argument. It does not attempt to present the usual *pot pourri* of information about the earth and its atmosphere which has

vii

traditionally been termed 'physical geography', but restricts itself to the identification and analysis of some of the more important systematic relation-ships with which modern physical geographers are concerned. Nor does it attempt to deal exhaustively with all the subject matter of this field. Its aim is simply to propagate the development of an up-to-date, relevant and intellectually-worthwhile approach to the discipline. It is possible that ultimately the real value of this book may lie in the intellectual stimulation it provides to view traditional material in physical geography in a new light.

Physical geography is facing an ever-deepening dilemma. On the one hand it has, partly for historical reasons, become responsible for both advanced research and elementary teaching in many earth science subjects (geomorpho-logy being an example of the former, meteorology of the latter); and, on the other hand, it is required to continue to play a relevant role with respect to a human geography which is becoming ever more economically and socially oriented. In the past it has often been possible to give at least the illusion of fulfilling both roles efficiently, but the increasing technical demands made on those who both research into and teach the earth sciences, and the increas-ing preoccupation of human geographers with spatial socio-economic matters, is making the position of the physical geographer more and more difficult. However, with, as we write, the mounting and undoubtedly justified concern about environmental destruction, disruption and pollution, and the growing number of examples of 'environmental backlash', it seems to us to be essential for geographers of all kinds to retain and reinforce their understanding of the manner in which the physical processes of this planet operate, and of the nature and complexity of the linkages between the inanimate and animate sectors of the real world.

During the present century not only the definition of physical geography (as distinct, in some cases, from 'earth science') but its position within geography as a whole, has changed drastically. This is a reflection of the changes we have mentioned in the whole human side of the subject, which have led to a marked tendency to emphasize purely spatial or socio-economic factors. We suggest that geography of any kind which becomes restricted to considerations of featureless plains, or their equivalent, is taking a dangerously unrealistic line and one, moreover, which is out of step with contemporary experience and need. The indications seem to us to necessitate spatial studies which not only include as many and as varied aspects of reality as possible but succeed in demonstrating a true understanding of the manner in which each of those aspects functions. This means that we consider it is not only valid but essential for geography to continue to possess a 'physical basis'. We do not, however, imply by this the classic man/land approach of the earlier part of this century, which was and is responsible for regional studies

that work from geology, to climate, to soils, to vegetation, to human settle-
ment. Rather, we consider that the 'physical basis' lies in those processes and
characteristics of the physical world which impinge directly upon, and
interact (or frequently conflict) with, those of socio-economic systems. We
hope that this book will provide an introduction to this view of the physical
basis of geography.

In addition, we have endeavoured to indicate to physical geographers how
far the results of traditional, non-systems-oriented studies can be seen to fit
the aims and requirements of systems theory. In this, as with all pioneers,
we are clearly laying ourselves open to the charge of placing old wine in new
bottles. We accept that there can be no genuine advance in our understanding
of the nature and operation of physical processes if we merely replace 'youth,
maturity and old age' by 'input, throughput and output'. However, by
attempting to recast a wide body of existing information within a systems
framework we hope to have pointed out not only the gaps in our current
knowledge of the systems and processes, but also the kinds of research which
are needed before we can genuinely evaluate the relevance of systems theory
to physical geography.

This book has, then, a dual purpose. At one level it attempts to present a
view of the landscape and its processes in terms which we feel are relevant to
the student of human geography and endeavours to indicate the ways in which
socio-economic and physical systems interlock and interact. At another, it
tries to show how far our knowledge of the physical world and its processes
are compatible with the ideas of systems theory and to demonstrate the
areas in which research might profitably be concentrated.

Richard J. Chorley,
Sidney Sussex College,
Cambridge,
England.

Barbara A. Kennedy,
New Hall,
Cambridge,
England.

Acknowledgements

The authors would like to thank the following publishers, editors, organisations, and individuals for permission to reproduce figures, tables and poems; and for permission to utilise published data. Figures refer to this book.

PUBLISHERS. Academic Press, New York: for 5.23 and 5.24 from Chorofas, *Systems + Simulations*, 1965; 8.3 from Murphy, *Adaptive Processes in Economic Systems*, 1965; 3.8, 3.12, 3.13 from Miller, *Advances in Geophysics*, 1965; 5.1, 5.14 from Court, *Advances in Geophysics*, 1952; 7.2, 7.14 from Holloway, *Advances in Geophysics*, 1958. Edward Arnold: for 1.3, 8.1, 8.2, 8.26 from Chorley, *Progress in Geography III*, 1971; 5.27 from Chorley and Haggett, *Network Analysis in Geography*, 1969; 6.32 from King, *Beaches and Coasts*, 2nd edition, 1971. Cambridge University Press: for 4.17, 6.11, 7.23 from volumes 98 and 106 of the *Geological Magazine*. Colston Research Society: for 1.2 by Chorley, from Chisholm (Ed.), *Regional Forecasting* (Butterworth). Columbia University Press: for 5.17, 5.18 Drawn from data in Gumbell, *Statistics of Extremes*, 1958. Elsevier, Amsterdam: for 8.18 from J. C. Ingle, Jr., *The Movement of Beach Sand*, 1966; and F. P. Agterberg, for 7.6 taken from *Earth Sciences Review III*, 1967. Freeman and Co., San Francisco: for 5.7b, 5.22, 5.26, 5.31, 5.33, 5.34, 6.4, 6.17, 6.18, 6.19, 7.7, 7.21 from Leopold, Wolman and Miller, *Fluvial Processes in Geomorphology*, 1964. Harcourt Brace Javonovitch, New York: for 6.3, 6.13 from Van Court Hare, *Systems Analysis: A Diagnostic Approach*, 1967. Harper & Row, New York: for 3.6 from Gates, *Energy Exchange in the Biosphere*, 1962. Harvard University Press: for 5.4 from Conrad and Pollack, *Methods in Climatology*, 1950; 8.22, 8.23 from Maass *et al.*, *Design of Water-Resource Systems*, 1962, Longmans and S. Gregory: for 5.5 from *Statistical Methods and the Geographer*, 1963. Macmillan: for 8.22, 8.23 from Maass *et al.*, *Design of Water—Resource Systems*, 1962. Macmillan Education: for 8.24, 8.25 from Hufschmidt and Fiering, *Simulation Techniques for Design of Water Resource Systems*, 1967. McGraw Hill, New York: for 2.3a, 2.32, 4.18 from Krumbein and Graybill, *Introduction to Statistical Methods in Geology*; 2.17 from Strahler in Ven te Chow, *Handbook of Applied Hydrology*, 1964; 3.22 from Kittredge, *Forest Influences*, 1964; 8.11 from Linsley, Kohler and Paulhus, *Applied Hydrology*, 1949. Methuen: for 1.8, 5.12, 7.9 from Barry and Chorley, *Atmosphere, Weather and Climate*; 3.14, 3.32, 6.7, 7.2, 8.19 from Chorley and Haggett, *Models in Geography*; 3.16, 3.26, 3.30, 3.33, 3.34, 3.35, 7.11, 8.9 from Chorley (Ed.), *Water, Earth and Man*. Oliver & Boyd: for 3.10, 3.21 from Sukachev and Dylis, *Fundamentals of Forest Biogeocoenology*, 1968.

Prentice-Hall Inc., Englewood Cliffs, N.J.: for 1.5, 7.5 from F. E. Croxton and D. J. Cowden, *Applied General Statisics*, 1939; 5.28 from C. Gordan, *System Simulation*, 1969. University of Chicago Press: for 2.7 from E. C. Olsen and R. L. Mills, *Morphological Integration*, 1958; for 3.3, 3.6, 3.7, 3.24, 3.25 from W. D. Sellers, *Physical Climatology*, 1953. Ronald Press, New York: for 8.12, 8.13 from L. B. Leopold and T. Maddock, Jnr., *The Flood Control Controversy*, 1954. Wiley, New York: for 1.9 from Strahler, *Introduction to Physical Geography*, 3rd edition, 1969; 2.3b, 2.6 from Ezekiel and Fox, *Methods of Correlation and Regression Analysis*; 7.1 from Mesarovic, *Views on General Systems Theory*, 1964; 7.4, 7.15, 7.34, 7.37 from Harbaugh and Merriam, *Computer Applications in Stratigraphic Analysis*, 1968; and two poems by Kenneth Boulding. American Geophysical Union: for 5.3.
EDITORS.   from Putz, *Transactions AGU*, **33**, 68, 1952; 5.13 from Stidd, *Transactions AGU*, **34**, 31–34, 1953; 7.8 from Leopold, *Transactions AGU*, **32**, 349–351, 1951; 8.14, 8.15 from Langbein and Schumm, *Transactions AGU*, **39**, 1076–1084, 1958; 8.17 from Stevens, *Transactions AGU*, Part II, 655, 1938. American Scientist: for 7.10 from Dorf, *Climatic Changes of the Past and Present*, 1960. *American Journal of Science:* for 6.6 from Carlston, **263**, 864–85, 1965; 6.9 6.10, 6.15 from Strahler, **248**, 1950; 6.25 from Langbein and Leopold, **262**, 1964; 7.22 from Schumm, **256**, 1956; 7.26 from Ruhe, **250**, 1952. Association of American Geographers: for 7.19 from the *Annals*, **50**, 1960. American Geographical Society: for permission to reprint from the *Geographical Review* 5.19 from Court, 1953; 8.5 from Rooney, 1967; 8.7 from Terjung, 1970; 8.10 from Burton-Kates, 1964. *Acta Sociologica*: for 6.30 from Carlssen, **II**, 1968. Four Corners Geological Society: for 7.31 from Pollack, *Guidebook*, 1969. Geological Society of America: for permission to reprint from the *Bulletin*, 2.18, 6.23, 6.24, after Strahler, **63**, 1952; 6.21 from Horton, **56**, 1945; 6.28, 7.20 from Schumm, **67**, 1956; 8.16 from Strahler, **69**, 1958. International Association of Scientific Hydrology: for 6.20 from Kirkby and Chorley, *Bulletin*, **12** (3), 1967; for 7.28 from Smart, Surkan and Considine, *Symposium on River Morphology Berne*, 1967; IASH Publication No. 75. Institute of British Geographers: for 2.11 from *Transactions* No. 37, 1965; for 7.24 from *Transactions* No. 18, 1952; for 7.32 from Special Publication No. 3, 1971. Institution of Water Engineers: for 3.18 derived from data in their *Journal*, Penman, **4**, 1950. *Journal of Geology*. University of Chicago: for 2.24, 4.8, 7.27 from Melton, **66**, 1958; 5.31, 6.14 from Wolman and Miller, **68**, 1960; 6.16 after Shreve, **75**, 1967; 7.25 from Welch, **78**, 1970; 7.33 after Culling, **71**, 1963. *Journal of Applied Meteorology:* for 5.7a from Hershfield, 1962; 5.25 from Feyerherm and Bark, 1965. *Journal of Geophysical Research*: for 5.15, 5.16 from Hershfield and Kohler, **65**, 1960. North Holland Publishing Company, Amsterdam: for permission to use figures from the *Journal of Hydrology*; 3.20 from Lewis and Burgy, 1964; 5.20 after

Rodda, 1967; 5.21 after Reich, 1963; 5.25 after Feyerherm and Bark, 1965; 5.35 after McPherson and Rannie, 1969. *Journal of Meteorology*: for 5.2 from Court, **8,** 1951. B JAP: 3.9 derived from data by H. L. Penman in **2,** 195, *Journal of Sedimentary Petrology* (Society of Economic Paleontologists and Mineralogists): for 2.2, 2.30, 2.31 after Krumbein, 1959. *Natural Hazard Research*: for 8.4 after Burton, Kates and White, 1968; 8.6 after Kates, 1970. *Quarterly Journal of the Royal Meteorological Society*: for 5.8 by Gregory, 1957. *Revue de Geomorphologie Dynamique*: for 5–11a derived from data by Caine, 1968. *Tellus* for 3.19 derived from data by Högstrom, 1968. *Weather* for 7.13 from Lewis, **15,** 1960.

ORGANISATIONS. Office of Naval Research and the authors: for 1.6, 4.2, 4.5, 4.6, 4.23, 5.29a derived from data by M. A. Melton, Dept. of Geology, Columbia University: in *Project* 389–042; *Technical Report* 11; 2.5 derived from data by M. Morisawa, Dept. of Geology, State University of New York in *Technical Report* 20, 1959. State Geological Survey, University of Kansas: for 7.35, 7.36 from Bonham-Carter and Harbaugh, 1968. University of North Carolina Press: for 2.9 after Blalock, *Causal Inferences*, 1964. U.S. Army Coastal Engineering Research Center: for 4.7, 4.21 derived from data by Harrison and Krumbein. USGS: for 2.12, 5.30 after Leopold and Maddock, 1953; 5.29b after Carlston, 1963; 6.5 after Wolman, 1955. Water Resources Research: for 8.20 after Fiering, 1965, **I,** 41–61. Dr. J. Amorocho.

INDIVIDUALS. for 3.31 from *A Tentative Glossary of Parametric Hydrology*, 1965. J. Harbaugh and F. Preston: for 4.19, 7.18 from a *Short Course and Symposium on Computers and Computer Applications in Mining and Exploration*, University of Arizona. A. D. Howard: for 7.29, from Department of Geography. The John Hopkins University, Baltimore: J. Klir and M. Valach for 6.29 from *Cybernetic Modelling*, 1967, by courtesy of Iliffe Books Ltd. J. Raynor: for 7.16, 7.17 from Department of Geography, Ohio State University. B. Sprunt: for 7.30, from Department of Geography, Portsmouth Polytechnic. W. Tobler: for 4.20 from "Geographical Filters and their Inverses," in *Geographical Analysis*, 1968. J. Towler: for 2.28, 4.10, 4.15, 6.8, from Department of Geography, University of Cambridge. M. Woldenberg of the Graduate School of Design, Harvard: for 1.1, from "Geography and Properties of Surfaces," *Harvard Papers in Theoretical Geography I*, 95–184.

The authors would like to thank the cartographic staff of the Department of Geography at Cambridge University for producing many of the figures; in particular Mr. M. Young, Mr C. Lewis and Mr R. Blackmore; and finally, Dr Adrian Harvey of Liverpool University who worked so hard on the index.

Publisher's note: Every attempt has been made to identify original sources of the illustrations but the Publishers would be grateful to hear of any errors or omissions, which will be acknowledged in future editions of the book.

Contents

List of Statistical Symbols

N.B. Many of these symbols are used at other points in the text to designate quite different qualities, but it should be clear from the context which definition is involved.

Greek letters

ϕ (Phi) Entropy.

Σ (Sigma: upper case) Summation. $\sum_{i=1}^{N} X_i$ means: add up the N different values assumed by variable X, from $X_i = $ 1st value, to $X_N = $ last value.

σ (Sigma: lower case) Standard deviation or 'root mean square' of a distribution. The square root of σ^2, the variance. Calculated, for a population, as:

$$\sigma = \sqrt{\sum_{i=1}^{N} [(X_i - \bar{X})^2]/N}$$

and, for a sample, as:

$$\sigma = \sqrt{\sum_{i=1}^{N} [(X_i - \bar{X})^2]/(N - 1)}$$

σ_{Ys} (Sigma sub $_{Ys}$) The standard error ('standard deviation of the sample means of variable Y') of the regression line $Y_C = a + bX$.

σ^2 (Sigma squared) Variance of a distribution. The square of the standard deviation (σ).

$\hat{\sigma}^2$ (Sigma carat squared) Variance of k samples about $\bar{\bar{X}}$:

$$\hat{\sigma}^2 = \sum_{i=1}^{km} [(X_i - \bar{\bar{X}})^2]/(km - 1)$$

$\hat{\sigma}_k^2$ (Sigma carat squared, sub k) Variance between k sample means:

$$\hat{\sigma}_k^2 = \sum_{j=1}^{k} [(\bar{X}_j - \bar{\bar{X}})^2]/(k - 1)$$

$\hat{\sigma}_m^2$ (Sigma carat squared, sub m) Variance within k samples:

$$\hat{\sigma}_m^2 = \sum_{j=1}^{k} \left[\sum_{i=1}^{m} [(X_i - \bar{X}_j)^2] \right] /(N - k)$$

Roman letters

a The intercept of a regression line i.e. Y_c, when $X = 0$; X_c when $Y = 0$.

b The slope of a regression line.

C The statistic for Cochran's Test of Equality of Variances.

e The base of natural or Naperian logarithms (2.71828).

E A random variable; the error term, unexplained variance, 'noise'.

F Snedecor's Variance Ratio statistic.

H Number of occurrences of one event of magnitude X in N time intervals

H_0 The null hypothesis of a statistical test.

k Number of samples.

m Number of observations within one sample; km = total number of observations = N.

n An exponent.

N Number of observations or variates.

N–1, N–2 etc. Number of degrees of freedom.

p The probability of an event occurring. p may vary between 0·0 and 1·0.

p_{ij} The transition probability of an event 'i' being followed by an event 'j'.

q The probability of non-occurrence of an event. $q = (1·0 - p)$ and may vary between 1·0 and 0·0.

Q_{25} The first quartile point: 25% of the observations are of lesser magnitude than Q_{25}.

Q_{75} The third quartile point: 75% of the observations are of lesser magnitude than Q_{75}.

r The coefficient of correlation: may vary from +1·0 to −1·0.

r^2 The coefficient of determination: may vary from 0·0 to +1·0.

R The coefficient of multiple correlation: may vary from +1·0 to −1·0.

R^2 The coefficient of multiple determination: may vary from 0·0 to +1·0.

t 'Student's' t statistic.

T The mean return period of an event of magnitude H. $T = N/H$ or $T = 1·0/p$.

U, V Spatial coordinates fitted by the method of least squares to a spatial distribution of values of Z to produce a three-dimensional trend surface.

u The ordinate of the normal curve: the distance of one observation from the mean, in units of standard deviations:
$$u = X_i - \bar{X}/\sigma$$

V_c Coefficient of variation: $V_c = \dfrac{\sigma}{\bar{X}} \cdot \dfrac{100}{1}$

V_r Relative variability: $V_r = \left[\sum\limits_{i=1}^{N} (X_i - \bar{X})/N \right] \cdot [100/\bar{X}]$

X A variable; an independent variable in any form of correlation analysis.

X_C A value of X lying on the best-fit regression line, $X_C = a + bY$.

X_i The ith value assumed by variable X: one observation or variable.

$\bar{X}$ The arithmetic mean of the distribution of X, derived from one sample:

$$\bar{X} = \sum\limits_{i=1}^{N} X_i/N$$

$\bar{X}_j$ The jth sample mean of a set of $j = 1 \ldots$ to k samples.

$\bar{\bar{X}}$ The arithmetic mean of the distribution of X, derived by pooling k samples of size m:

$$\bar{\bar{X}} = \sum\limits_{i=1}^{km} X_i/km$$

$\check{X}$ The median value or second quartile point (Q_{50}) of the distribution of X. Divides the X_i into two groups of equal number.

$\hat{X}$ The modal or most commonly occurring value in the distribution of X.

Y The dependent variable in simple or multiple regression; an independent variable in trend surface analysis.

Y_C A value of Y lying on the best-fit regression line $Y_C = a + bX$.

Y_i The ith value assumed by variable Y: one observation or variable.

$\bar{Y}$ The arithmetic mean of the distribution of Y:

$$\bar{Y} = \sum\limits_{i=1}^{N} Y_i/N$$

Z A variable; the dependent variable in trend surface analysis.

Z_i The ith value assumed by variable Z: one observation or variable.

Z_C A value of Z lying on the best-fit trend surface.

1: *Systems*

1.1 The Complexity of Reality

The real world is immensely complex. Man reacts to this by trying, first, to isolate parts of reality—either in fact or in theory—and, second, to investigate how the parts operate under simplified conditions. Although intellectually necessary, this decomposition of the real world into simplified structures is an entirely subjective product of the mind of the investigator. Moreover, the ultimate aim of such an investigation must be the linkage of the simplified structures so identified with others, on the same or differing scales of space or time.

The real world is continuous. Isolated structures are therefore subjective and artificial portions of reality, and the biggest initial problem is the identification and separation of *meaningful* sections of the real world. On the one hand, every section or structure must be sufficiently complex to possess a high degree of internal coherence, so that its study will yield significant and useful results; on the other, every section must be simple enough for comprehension and investigation.

All such studies have, as their central theme, the analysis of the manner in which the components of each section of the real world are internally structured (i.e. organised, linked, or related) and, further, how each section links to other structures. These structures are commonly termed SYSTEMS.

Clearly systems can be identified at all scales of magnitude and with all degrees of complexity. We can, therefore, consider that 'reality' is a hierarchy of organised systems. Indeed, it may be that such a view of the real world—as an interlocking set of systems—will provide a greater impetus to the unification of all sciences than will their analysis in terms of physical and chemical attributes. The real world, then, can be viewed as comprising sets of interlinked systems at various scales and of varying complexity, which are nested into each other to form a systems hierarchy (i.e. of subsystems, systems, super-systems, etc.).

1.2 Systems

A system is a structured set of objects and/or attributes. These objects and attributes consist of components or variables (i.e. phenomena which are free to assume variable magnitudes) that exhibit discernible relationships

1

with one another and operate together as a complex whole, according to some observed pattern.

Interdependence of parts is a diagnostic property of systems. As every system is made up of a unique set of parts, related in a specific manner, it is frequently said that such structures are *gestalt:* that is, the whole is greater than the sum of the parts. This is, however, rather a misleading phrase, as it implies, incorrectly, that the inherent character of a system involves the addition of its parts plus some new factor that appears when summation occurs. This interrelationship of the parts of a system may be expressed either as a correlation between the observed magnitudes of the objects and attributes of which it is composed, or in the harmony of response which is exhibited by one component when mass or energy from another component of the system is directed into it.

Just as the isolation of individual systems as objects of study represents one way of classifying the phenomena of the real world, so there are several ways in which systems themselves can be classified. The most common division is the *functional* one, into isolated, closed and open systems.

1 Isolated systems are assumed to have boundaries closed to the import and export of both mass and energy. Such structures, clearly, occur more frequently in the laboratory than in the real world. It is, however, often convenient to assume that some particular system is isolated, in order to determine the degree to which external variables really influence its behaviour. This may be done by noting the discrepancies between the observed behaviour of the system variables and that which may be ascribed solely to their internal relationships or the mass and energy flows between them.

2 Closed systems are assumed to have boundaries which prevent the import and export of mass, but not of energy. The earth together with its atmosphere, represents a system which is, to all intents and purposes, closed, exchanging energy but not mass with outer space. Closed systems of this type occur quite frequently in the real world.

3 Open systems, on the other hand, are characterised by an exchange of both mass and energy with their surroundings. The components of such systems and the interrelationships between them tend to become adjusted, so that there is a steady output of mass and energy (or *information*) which is equal to the input. Such adjustment is termed *self-regulation* operating to produce a *steady state*. The majority of natural systems are open. For example, a drainage basin receives energy from sunlight, falling precipitation and the elevation of the land surface, together with mass in the form of water, salts and the products of disintegration of the underlying bedrock: these inputs provide the basis for the outputs of heat, water, organic and inorganic debris into the atmosphere or the sea.

One of the outstanding features of both closed and open systems is the manner in which the throughput of energy tends to produce and maintain a discernible organisation, characterised by hierarchical differentiation. This feature is common to both animate and inanimate systems. Fig. 1.1. illustrates how a system of stream channels forms such a hierarchy, with smaller headwater tributaries feeding larger channels of higher order in the hierarchy. The figure further illustrates how a given geometrical hierarchy may be theoretically ordered in a number of different ways with major orders continuing right up to the headwater (Fig. 1.1a); with fingertip tributaries being designated order 1 and order 2 channels being formed by the joining of two order 1 tributaries, and so on (Fig. 1.1b); or with each channel segment having an order determined by the number of order 1 fingertip tributaries which ultimately feed it (Fig. 1.1c).

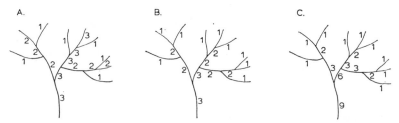

Fig. 1.1   Three methods of designating the hierarchy of stream channels present in a drainage basin. (After Woldenberg)

(a) Horton's stream orders.
(b) Strahler's stream orders.
(c) Shreve's stream magnitudes.

The development of hierarchies of subsystems within closed and open systems is sometimes brought about by the need for structures of differing magnitudes and complexities in order to perform different, and often incompatible, tasks. This need is reinforced by the existence of *thresholds*.

The second major way of classifying systems is on the basis of their internal complexity: this, then, is a *structural* classification. In rough outline, the complexity of natural systems may be thought to increase in the following stages:

1 *Morphological systems*. These consist purely of the network of structural relationships between the constituent parts of systems.
2 *Cascading systems*. These systems are defined by the path followed by throughputs of energy or mass.
3 *Process-responses systems*. These represent the linkage of at least one morphological and one cascading system, so that the process-response system demonstrates the manner in which *form* is related to *process*.

4 *Control systems (Transducers).* These are process-response systems in which the key components are controlled by some intelligence. This control causes the system to operate in some manner determined by the intelligence.

5 *Self-maintaining systems.* These represent the lowest form of life (e.g. cells).

6 *Plants:* as living structures.

7 *Animals:* as living structures.

8 *Ecosystems.* These systems are made up of plants, animals and their inanimate environment.

9 *Man.*

10 *Social systems.*

11 *Human ecosystems.* These systems represent the interlocking of social systems with ecosystems.

At each of these levels of complexity, one can identify systems of differing scale. More important, however, is the division between the fourth and fifth levels of complexity: that is, between inanimate and animate (or self-re-producing) systems. The link between these categories is provided by the common tendency for the production of hierarchical levels of organisation, as the result of the throughput of energy and/or mass.

1.3 Systems in Physical Geography

Geography is concerned with medium-scale systems; those lying somewhere between the scales of the atom and the universe. Physical geography is concerned primarily with four types of systems, each exhibiting distinct but

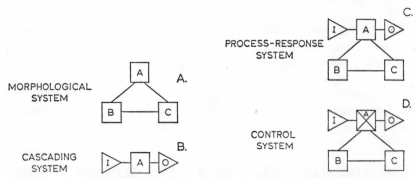

Fig. 1.2   Schematic illustrations of the four types of systems identified in this book

complementary properties, and forming a progressive sequence to higher levels of integration and sophistication. These are the morphological, cascading, process-response and control systems identifiable at the surface of the earth (Fig. 1.2). A secondary consideration is the interaction of these relatively low-complexity systems with the higher-level structures of plants, animals and human societies. The latter systems are, however, only of relevance to the physical geographer in certain cases:

1 If they represent morphological features at a geographical scale (e.g. woods).
2 If they play a part in cascading systems (e.g. the role of vegetation in the energy budget or hydrological cycle).
3 If they provide the intelligence which intervenes in, and controls, natural process-response systems (e.g. the removal of vegetation; weather control, etc.).

Having identified the structural classes of systems which are of primary concern to the physical geographer, it is necessary to consider their characteristics in greater detail.

A. MORPHOLOGICAL SYSTEMS (Fig. 1.3*a*)

These comprise the morphological or formal instantaneous physical properties integrated to form a recognisable operational part of physical reality, the strength and direction of their connectivity being commonly revealed by correlation analysis. So, for example, the morphological properties of a beach system might include such parameters as beach slope, mean grain size, range of grain sizes, beach firmness, and the like. The relationships between these parameters can be expressed by a web of correlations, and it is usual to interpret the operational efficiency of such a system in terms of the degree to which these morphological parameters are related. Thus, although these morphological variables can be identified individually, their interrelationships are often indicative of the degree to which their dynamic properties are related. An important feature of such systems is the role played by *feedback loops,* particularly negative ones, in governing the general morphological changes which follow changes in individual variables. These systems are in some ways analogous to mechanical structures, where an externally applied stress is released by a strain (i.e. a readjustment of the associated variables, to produce a new equilibrium).

B. CASCADING SYSTEMS (Fig. 1.3*b*)

These are composed of a chain of subsystems, often characterised by thresholds, having both spatial magnitude and geographical location, which are dynamically linked by a cascade of mass or energy. In this cascade the mass or energy output from one subsystem becomes the input for the adjacent locational subsystem. In this way, for example, the output of debris and surface runoff

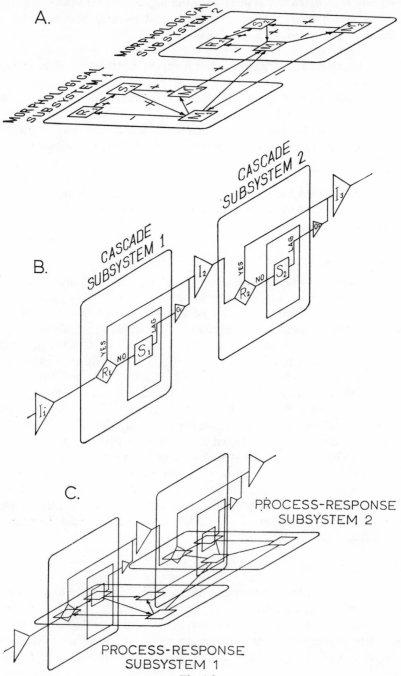

A.

MORPHOLOGICAL SUBSYSTEM 1

MORPHOLOGICAL SUBSYSTEM 2

B.

CASCADE SUBSYSTEM 1

CASCADE SUBSYSTEM 2

C.

PROCESS-RESPONSE SUBSYSTEM 2

PROCESS-RESPONSE SUBSYSTEM 1

Fig. 1.3

from the slope subsystem becomes part of the input of water and debris into the stream channel subsystem and partly governs its operation. Such cascading systems may be simply viewed in terms of basic *canonical structures*, in which an input into a subsystem is acted upon by a *decision regulator* such that some mass or energy may be diverted into a subsystem store and the remainder throughput to become subsystem output into adjacent subsystems. Cascading systems in physical geography vary markedly in magnitude, from the basic solar energy cascade, to the basin hydrological cycle cascade, and to the cascading subsystem formed by a wave moving from the deep water to the swash zone subsystem.

In studying such systems, the whole emphasis is placed on the relationships between input and output and it is convenient here to draw attention to three ways of analysing cascades (Fig. 1.4).

(*a*) White box. An attempt is made to identify and analyse as many of the storages, flows, etc., as possible in order to obtain the most detailed knowledge concerning the manner in which the internal structure of the system produces a given output in response to a given input.

(*b*) Grey box. This involves a partial view of the system, in which interest is centred on a limited number of subsystems, the internal operations of which are not considered.

(*c*) Black box. The whole system is treated as a unit, without any consideration of its internal structure and attention is directed solely to the character of the outputs which result from identified inputs.

Fig. 1.3 Diagrams showing the more detailed character of morphological systems, cascading systems, and their intersection to form process-response systems. (From Chorley, 1971)

(*a*) A morphological system containing two subsystems composed of eight morphological variables (R_1, S_1, M_1, M_1^*, R_2, S_2, M_2, M_2^*), linked to form a correlation structure.

(*b*) A cascading system containing two subsystems connected by inputs and outputs (I_1, I_2, I_3) which are disposed by regulators (R_1, R_2) into storages (S_1, S_2) from which lagged outputs (O_1, O_2) occur.

(*c*) The above morphological and cascading systems linked to form a process-response system by the sharing of two regulators (R_1, R_2) and two storages (S_1, S_2) which simultaneously perform the roles of cascade components and morphological variables.

BLACK   BOX

GREY   BOX

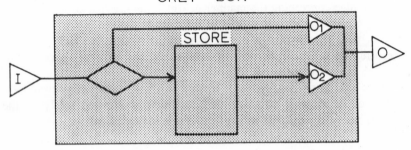

WHITE BOX

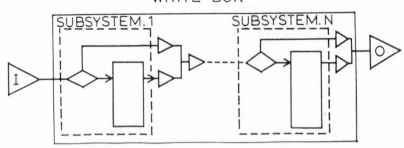

Fig. 1.4   Three ways of viewing cascading systems—black box, grey box and white box systems

C. PROCESS-RESPONSE SYSTEMS (Fig. 1.3c)

These are formed by the intersection of morphological and cascading systems. The links between the two types of system are provided by morphological states, which either coincide, or are closely correlated, with storages or regulators embedded in the cascading systems. This integration is generally produced either by a sharing of certain variables by the two types of systems (as when, for example, infiltration capacity is both a morphological property of a slope system and a decision regulator in the basin hydrological cycle cascading system), or when a high level of correlation exists between a variable in the cascading subsystem and one or more in the associated morphological system. These linkages between the systems commonly involve negative-feedback processes so that the morphological responses to changes in the energy cascade of such a system are dominated by self-regulatory processes. This type of negative-feedback relationship is dominant in physical processes, and one of the problems when attempting to produce a system which includes both physical and human phenomena is that many aspects of socio-economic systems are governed by positive feedback, as when, for example, there is a 'snowballing' effect of capital accumulation or increasing locational advantage. It is thus possible to view process-response systems as consisting of cascading and morphological components which mutually adjust themselves to changing input-output relationships. The time required for such adjustment to a new equilibrium (i.e. the *relaxation time*) being dependent on such considerations as the amount and direction of change in the energy cascade and the number of links between the variables.

In isolating process-response systems, the emphasis is placed on identifying the relationships between a process and the forms resulting from it. The states of storages in cascading systems are particularly important system links. For example, if the infiltration cascade fills the soil moisture storage, it leads to overland flow linking this with the streamflow cascade. In this way, the morphological systems of the slope and the stream channel become linked by the cascade component of surface runoff.

D. CONTROL SYSTEMS (Fig. 1.2d)

The fourth level of systems integration is represented by the control system. When one examines the structure of physical process-response systems it becomes apparent that certain key variables or *valves* (commonly involving the decision regulators) are those wherein intelligence can conveniently intervene to produce operational changes in the distribution of energy and mass within cascading systems, and consequently to bring about changes in the equilibrium relationships involving the morphological variables linked with them in the process-response systems. Thus, for example, man-induced changes in drainage basin infiltration capacity or sand movement on a beach can produce, either knowingly or inadvertently, considerable changes in drainage density and beach geometry. Where such human intervention is not inadvertent, perverse or isolated, but results from the operation of a coordinated spatial decision-making system, the resulting interaction of this decision-making system and the physical process-response system produces a *geographical control system*. These systems can be visualised as three-dimensional structures in which the very complicated flows and relationships forming the socio-economic spatial decision-making systems interpenetrate the physical process-response systems at the control valves (see Chapter 8).

1.4 The Structure of Systems

The diagnostic feature of any system lies in the arrangement of its components and it is this aspect which dominates the following characteristics, by which a system is defined:

1 *Size*. The size of a system is determined by the number of variables of which it is composed. Of course, this means that each variable is treated as a black box, whereas in reality, each is a subsystem made up of component variables in its own right. Where the system is composed of completely interlocking variables (i.e. each one is related to all the others) then its complexity and size are expressed in terms of the *phase space*, or number of variables. Fig. 1.5 illustrates data relating to a simple 2-phase-space system the line representing the generalised relationship between the variables.

2 *Correlations*. The correlation between pairs of variables in a system is expressed by the closeness with which a generalised relationship fits the data. This generalised relationship is a mathematical regression line (such as that shown in Fig. 1.5*a*) which is usually derived by considering a *sample* of all possible values which two of the variables in the system may attain. (This topic is discussed at length in Chapter 2.)

The aspects of correlation which specify the nature of the system structure are:

(*a*) The strength. This is an expression of the degree to which the data points coincide with the fitted regression line and it is described by the *coefficient of correlation* (r) which approaches unity as the points coincide more and more exactly with the line. The degree to which the points fail to coincide with the line is termed *noise:* this is an analogy with a radio signal, in that noise obscures the relationship between the system components.

A.

$$Y_C = -137.22 + 3.777 X$$

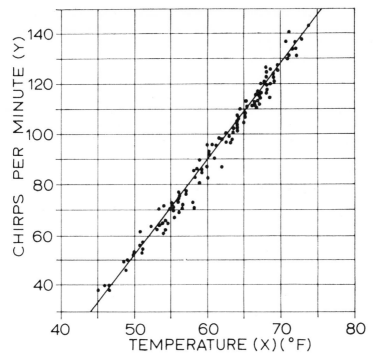

B.

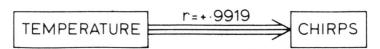

Fig. 1.5   The relation between cause and effect. (From Croxton and Cowden, 1948)
(*a*) A simple linear regression showing the relationship of number of cricket chirps per minute (*Y*) as a function of temperature (°F) (*X*)
(*b*) A schematic correlation linkage expressing the above relationship in symbolic form

(*b*) The sign. The positive or negative sign attached to the correlation coefficient indicates whether an increase in the magnitude of one variable is associated with an increase or a decrease in the value of the other.

(*c*) The sensitivity. This indicates the extent to which a change in the magnitude of one variable leads to a change in the other. In a sensitive relationship a small change in one variable will produce a large change in the other.

3

(*d*) The probability. Every correlation is based upon observations of the magnitude of the variables made at a limited number of sampled points. It is, therefore, quite possible that the apparent strength and sensitivity of the correlation may be due to the fact that, by chance, points have been sampled which produce a close scatter around a line of obvious trend. Consequently, further observations of the relationship might produce an increase in the scatter and a simultaneous decrease in the slope, or trend, of the line.

A relationship of high probability has a large value of 'r' and a steep slope: in other words, one can be reasonably certain that the continued collection of data will not substantially weaken the correlation which has already been inferred.

The relationship shown in Fig. 1.5*a* indicates a case of high, positive correlation, with relatively high sensitivity and a very high probability. Fig. 1.5*b* illustrates some of these features in symbolic form.

3 *Causality*. The direction of causality (shown by an arrow) indicates which is the *independent* (i.e. controlling) variable and which the *dependent* one, such that the latter will always and only change if the former does. This arrow is usually inserted as the result of experience of the way in which the system operates, or from common sense (e.g., in Fig 1.5 it is clear that varying temperature controls the rate of chirping and not vice versa!), but in complex systems some arrows are inserted to explain the way in which the system, as a whole, seems to be operating. This introduces us to such concepts as *feedback*, which will be treated in Chapter 1.5.

4 *Pattern*. Most systems which are identified as a useful basis for analysis actually possess more than two variables and the pattern of intercorrelations within them is one of their most diagnostic features. Where two or more variables have a joint effect on another, this composite entity is termed a *vector*. The effect of the vector may be one of simple *additivity*, when the composite effect is the same as the sum of the individual effects. More usually, however, *replication* (or redundancy) occurs, when the joint effect is less than the summed total: in this instance it is clear that one or more variables are repeating the information provided by others (i.e. that some overlap is occurring between the components of the vector). Occasionally the opposite effect, of *reinforcement*, is found, where the total effect is greater than that of the sum of the individuals and the effect of one variable is being heightened by the presence of another. A way of expressing the degree of interaction of variables in a vector is in terms of their number of *degrees of freedom*, which describes the number of completely free variables which would

be necessary to produce the observed range of possibilities. For example, the three British traffic lights (red, amber and green) produce 4 combinations, or signals; in New York, two lights (red and green) also produce 4 signals: red; green; red and green; no light showing. The system of three lights therefore can be said to have two degrees of freedom.

The most important aspects of system pattern are those directed linkages which will be treated under the subject of feedback.

5 *Externalities.* It is a feature of systems that one or more of their component variables are linked to external variables. The nature of this linkage is explained more fully in the chapter on process-response systems.

6 *Output.* The output of a system can be of two kinds. Firstly, the actual output of energy, mass, or information and, secondly, the changes of form, storage, structure, etc., resulting from a given input. In reality, however, this distinction is rather artificial, as both types of output represent changes in the storages within the system and, consequently, alterations in the intensity of subsequent throughputs.

1.5 The State of Systems

In an isolated system, with no energy input, there is a tendency for energy to be continuously redistributed throughout the system in a more and more random fashion. This quality of randomness or disorganisation is termed *entropy* and, as entropy increases through time, it represents a decrease in the amount of energy available for work, a levelling-down of differences within the system and a destruction of its hierarchical organisation. This process continues irreversibly until the isolated system reaches the static equilibrium of an undifferentiated structure, with a uniform distribution of energy and entropy at a maximum.

Open and closed systems are continuously importing energy and, as this increases the differentials of energy distribution and promotes hierarchical organisation and structure, this imported energy is considered to be *negative entropy*. The state of an open or a closed system cannot, therefore, be described with reference to static equilibrium. However, the interconnecting structure of such systems imparts the quality of *self-regulation* to them and this leads to a different type of equilibrium. Complex open and closed systems are composed of integrated feedback subsystems which operate in a self-regulatory manner and enrich the system by providing it with greater flexibility of response.

Feedback is the property of a system or subsystem such that, when change is introduced via one of the system variables, its transmission through the structure leads the effect of the change back to the initial variable, to give a circularity of action. Feedback can be either direct (Fig. 1.6*a*) or looped, by means of one or more variables (Fig. 1.6*b*). The most common type of feedback is *negative feedback* (Fig. 1.6*c*) where an externally-produced variation sets up a closed loop of change which has the effect of damping down or stabilising the effect of the original change. This situation is indicated by a loop with an odd number of negative correlation signs: for example, in Fig. 1.6*c*, an increase in B causes an increase in C; and increase in C causes a decrease in D; which leads to a decrease in B. This is a form of self-regulation which promotes *open system* or *dynamic equilibrium*. Negative feedback

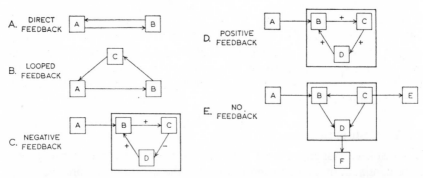

Fig. 1.6   Symbolic diagrams illustrating the various types of feedback relationships

systems and subsystems are extremely common in nature and are responsible for its conservative and self-repairing features. *Positive feedback* occurs when closed loops of variables reinforce the effect of externally induced changes, to give snowballing of the changes in the same direction as their initial action. Such loops have either an even number or no negative correlation coefficients (Fig. 1.6*d*). It is apparent that such positive feedback systems cannot operate unchecked and the limits are provided by individual variables which are unable to change indefinitely in one direction. These systems therefore operate in relatively short bursts of self-destructive activity. Of course many systems and subsystems show no pattern of feedback (Fig. 1.6*e*), being a linkage of independent (A), intermediate (B, C), dependent (D, E) and singly correlated (F) variables.

Although one can assess in general whether a system exhibits positive or negative feedback by noting the number and direction of links within it, it is most important to realise that the *magnitude* of the links is of even greater significance. That is to say, one extremely strong relationship—whether positive or negative—can prove all-important as far as the actual operation of the system is concerned and may lead to a real negative feedback situation where a positive one would be expected, or *vice versa*. It is also necessary to stress that the strength of individual linkages will vary not only between different, though similar, systems, but also within the same system at various times.

The dominantly negative feedback character of natural systems, or of key regulating subsystems within them, means that any change in the energy environment of the system will result in a change of the system variables which will, in turn, lead to a new system equilibrium. This tendency for self-regulatory change is called *dynamic homeostasis* and the equilibrium which results from the change in energy input is termed a *steady state* of operation. The effect of dynamic homeostasis is usually shown by high correlations between systems and subsystems with dominantly negative feedback and by the tendency for the observed values of individual variables to crowd around very characteristic magnitudes.

This simple self-regulation based upon negative feedback is complicated by the existence of both *secondary responses* and *thresholds*. Secondary responses are those which may eventually result from external changes of input and affect the operation and equilibrium of the system some time after the adjustment to the initial energy change. For example, a decrease in precipitation may cause stream channels to become wider and shallower, but a longer-term effect of the change in precipitation may be a decrease of vegetation cover, permitting more rapid storm runoff and leading to channel trenching. Thresholds involve drastic changes in state as they are traversed by the system, and a system which habitually transforms into radically different states as the result of traversing thresholds is termed *metastable*.

1.6 Input/Output Relationships

We have seen how a change in input may produce changes in system response, or output, and how this will tend to produce a new steady state, the output of which will consist either of an internal organisation characterised by high intercorrelations, or a steady output of energy or mass. The time taken by the system reorganisation to achieve a new steady state, following a change of input, is termed the *relaxation time*. The length of this time period, in which the retarded process of self-adjustment within a perturbed system leads to the attainment of a new equilibrium, is dependent on:

1 The states of the individual variables within the system. In cascading systems the state of storage variables is particularly important. For example, the fact that the soil moisture storage in the Lyn catchment was

already full when the storm input of August, 1952, took place was largely responsible for the very short relaxation time before flood output from the basin occurred.

2 The resistance to change of the individual variables in the system.

3 The complexity of the system. Usually the more complex the system (i.e., the greater the phase space and the more it is composed of coupled and interlocked negative feedback subsystems) the shorter the relaxation time tends to be, because of the number of possible combinations of equilibrium relationships. Because many reaction steps involve simultaneous multiple reactions throughout the system, a complex system usually possesses a whole spectrum of relaxation times, some of which are sequentially linked in a sophisticated manner.

4 The magnitude and direction of change. Generally, the rate of readjustment is directly related to the distance of the variable from its new equilibrium state.

Relaxation time differs widely in natural systems. Alterations of the channel geometry of alluvial streams following changes in precipitation, for example, may be virtually instantaneous, but changes in valley-side slope geometry or drainage density may take much longer. The relaxation time of a system is a major factor in determining the importance which the investigator places on historical evolution as a control of the present state of the system. It is clear that where the relaxation time is short, the timeless character of the system appears dominant and the investigator looks for equilibrium relationships: where it is long, the timebound or historical features preoccupy him, as he traces the changes in the system through a sequence of progressively more adjusted states. Different subsystems usually have very different relaxation times and natural systems commonly form an interlocking pattern of adjusted, partly-adjusted and poorly-adjusted subsystems. Some of them may be so well adjusted that they may be quite adequately analysed in terms of present energy inputs, whereas others may be so resistant to change that they are understandable only in terms of past processes. It is significant that the areas of greatest methodological conflict are often those relating to phenomena where information regarding relaxation times is lacking (e.g. slope studies in physical geography). This is even more pronounced where studies involve complexes of subsystems having different, but largely unknown, relaxation times (e.g. urban systems in human geography).

This picture of input, adjustment and output is, of course, very over-simplified when applied to most natural systems. This is because the relaxation time is confused by the occurrence of secondary responses with delayed relaxation times or, in some cases, by *overshooting*. The latter occurs when a relatively small input causes the system to cross a threshold and enter into a new economy. An example of this is when moisture changes in slope debris lead to a progressive loss of internal strength, until a time comes when a small addition of moisture decreases the strength to below the critical level of stability and catastrophic failure occurs, producing a new slope system with very different characteristics (Fig. 1.7). However, the major factor which complicates the pattern of outputs is the pattern of inputs. Most natural inputs (e.g. of solar energy, precipitation, wind strength, wave height, etc.) fluctuate in such a complex manner that they must be treated as if they possess a *random* element. Where input processes are partly random in character (i.e. *stochastic*) questions regarding system behaviour become probabilistic. For example:

1 What is the chance of the system receiving a given input at a given time? This question relates to the *magnitude and frequency* of input events, which will be treated in a later chapter. Fig. 1.8 gives the probability of occurrence of rainstorms of given amount and duration at Cleveland, Ohio: probability being stated in terms of the average recurrence interval, which is obtained from the average number of storms of a particular magnitude expected in a given time period.
2 What is the chance of a given input being followed by another of a particular magnitude? The answer to this question involves analysis in terms of a *Markov process* of the form:

$$y_{t+1} = a.y_t + bE$$

where: y_t = the magnitude of the occurrence at time t; y_{t+1} = the magnitude of the occurrence at time $t + 1$; a = a constant; b = a constant depending upon the range of values which y may have; and E = a random variable.

Because different components of one system may have differing relaxation times, it is common, when the size or nature of inputs from other systems is fluctuating rapidly, for the inertia of the system operation to filter out some responses or to blur them by coalescence with other responses.

The existence of the various system relaxation times means that if, as is common, external changes of input are fluctuating rapidly then the inertia of the system operation means that some responses are filtered out, or blurred by coalescence with other responses (e.g. as are the effects of small rainstorms on the runoff from a large catchment, when they are followed by heavy rainstorms).
3 What is the chance of a given input producing a given output in a given time? This follows immediately from the previous question and the answer depends upon the inputs with which it is associated in time. One

Systems

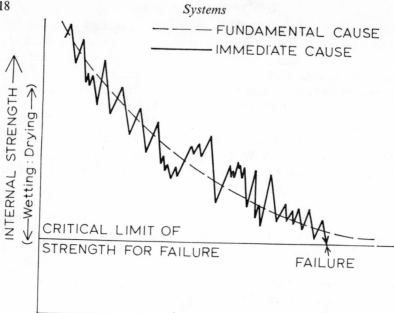

Fig. 1.7

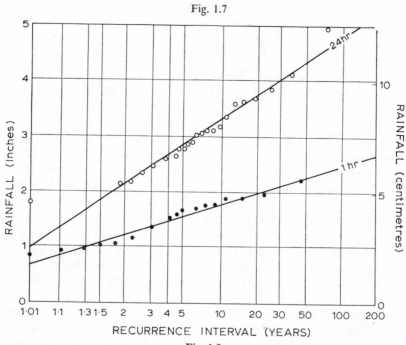

Fig. 1.8

way of treating this problem is to use *queueing theory*, whereby any queueing process can be defined in the following terms:

(*a*) The arrival pattern—i.e. the magnitude and frequency of the inputs.

(*b*) Service process—i.e. the service time and capacity. These determine the connection delay, or the time between input and output. Fig. 1.9 illustrates this in terms of the basin *lag* time between maximum rainfall intensity and maximum stream discharge for a drainage basin in Ohio.

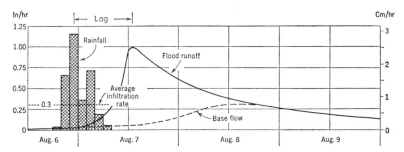

Fig. 1.9   A simple illustration of a lag between rainfall input and stream discharge output from a small drainage basin. The storm produced 6·3 inches of rain in 12 hours over the Sugar Creek drainage basin in Ohio (310 square miles), and of this some 3 inches contributed to the flood runoff the peak of which showed a basin lag of more than 24 hours after the time of maximum rainfall intensity. (After Hoyt and Langbein, and Strahler. From More, 1967)

(*c*) Queue discipline—i.e. what the rules for 'queue jumping' are. For example, some of the rainfall from one storm will contribute to soil water storage, thus ensuring that a portion of the rainfall from a later storm will by-pass the soil storage subsystem and reach the stream channels more rapidly, as surface runoff.

(*d*) Attrition rate—i.e. loss of input from the system before it forms part of the measured output. The evaporation of water from the surface of a drainage basin will deplete the streamflow output in comparison with the rainfall input.

The term random is commonly applied to an occurrence which is unpredictable with any degree of certainty for a given event; this contrasts

Fig. 1.7   The separation of a fundamental cause from an immediate cause of failure in material subject to a sequence of wetting and drying (causing decreases and increases of internal strength respectively), such that the repetition of this cycle leads to a progressive loss of internal strength (i.e. the fundamental cause of failure) until a final small wetting phase (i.e. the immediate cause) causes failure. (Partly based on Terzaghi)

Fig. 1.8   Rainfall/duration/frequency curves for Cleveland, Ohio (1902–1947), showing the average recurrence intervals of different amounts of rain falling during 1 hour and 1 day. (After Linsley and Franzini, 1955.) A more complete treatment of this type of analysis is given in Chapter 5

with a *deterministic* occurrence whose behaviour is uniquely predictable by mathematical functions. Where a random element is present in an otherwise apparently deterministic occurrence the departure from predictable regularity attributable to this random element is termed *noise*. However, statistical regularity may result from events which may be unpredictable *as individuals*, but predictable *as a group* when judged from a large number of observations. The science of *statistical mechanics* is concerned with the generation of such statistical laws (see Chapter 6.7). There have been many suggestions as to the 'causes' of randomness, some of them highly metaphysical, but from our point of view the following are of especial interest:

1 Indeterminacy. This relates to situations in which discernible physical laws apply, but ones which may be satisfied by a large number of possible combinations of values of the independent variables. An example of this in hydraulic geometry is given in Chapter 6.6.
2 Superimposition. The accumulation or superimposition of many deterministic influences into a complex tangle.
3 Partiality. Resulting from the fact that a system is never a truly isolated one and its behaviour is therefore linked, to a greater or lesser extent, with that of other systems, the interaction of which introduces apparently random effects in the system under observation.

A vital question, which will be examined more fully in a later chapter, is the way in which a closed or open system can change progressively through time, thereby introducing an historical element into its development. We need only say here that such changes are either due to progressive alterations in the system inputs, or to some progressive modification of the state or organisation of the system variables. For example, the cascade system of the basin hydrological cycle changes progressively throughout each storm because one of its important components—infiltration capacity—decreases in magnitude with time. This is akin to the longer-term progressive loss of debris strength—leading to slope failure—to which reference has already been made. However, more profound changes with time are characteristic of most natural systems: a major instance is the effect of the irreversible removal of debris upon the development of an erosional drainage basin.

REFERENCES

Ackoff, R. L. (1960), Systems, organization, and interdisciplinary research; *General Systems Yearbook*, **5**, 1–8.
Amorocho, J. and Hart, W. E. (1964), A critique of current methods in hydrologic systems investigation; *Transactions of the American Geophysical Union*, **45**, 307–321.

Anderson, J. (1969), On general systems theory and the concept of entropy in urban geography; *London School of Economics, Graduate Geography Department, Discussion Papers No. 31*, 17.

Ashby, W. Ross (1958), General systems theory as a new discipline; *General Systems Yearbook*, **3**, 1–6.

Ashby, W. Ross (1964), *An Introduction to Cybernetics* (Methuen, London), 295.

Beer, S. (1959), *Cybernetics and Management* (Wiley, New York), 214.

Blalock, H. M. (1964), *Causal Inferences in Nonexperimental Research* (University of North Carolina Press), 200.

Borchert, J. R. (1967), Geography and systems theory; In Cohen, S. B. (Ed.), *Problems and Trends in American Geography* (Basic Books, New York), 264–272.

Boulding, K. (1956), General systems theory—The skeleton of science; *General Systems Yearbook*, **1**, 11–17.

Chisholm, M. D. (1967), General systems theory and geography; *Transactions of the Institute of British Geographers*, No. 42, 45–52.

Chorley, R. J. (1962) Geomorphology and general systems theory; *U.S. Geological Survey, Professional Paper* 500-B, 10.

Chorley, R. J. (1967), Models in geomorphology; Ch. 3 in. *Models in Geography*, Ed. by Chorley, R. J. and Haggett, P. (Methuen, London), 57–96.

Chorley, R. J. (1971), The role and relations of physical geography; *Progress in Geography*, Vol. 3, 89–109.

Clarke, D. L. (1968), *Analytical Archaeology* (Methuen, London), 684.

Conacher, A. J. (1969), Open systems and dynamic equilibrium in geomorphology: A comment; *Australian Geographical Studies*, **7**, 153–158.

Cowan, T. A. (1963), On the very general character of equilibrium systems; *General Systems Yearbook*, **8**, 125–128.

Emery, F. E. (Ed.) (1969), *Systems Thinking* (Penguin, London), 398.

Gale, S. (1970), Simplicity isn't that simple; *Geographical Analysis*, **2**, 399–402.

Hall, A. D. and Fagen, R. E. (1956), Definition of system; *General Systems Yearbook*, **1**, 18–28.

Heal, D. W. (1968), Geography, general systems theory and common sense; *University of Newcastle-upon-Tyne, Department of Geography, Seminar Papers No. 3*, 25.

Howard, A. D. (1965), Geomorphological systems—equilibrium and dynamics; *American Journal of Science*, **263**, 302–312.

Kern, R. and Weisbrod, A. (Trans. D. McKie, 1967), (1964), *Thermodynamics for Geologists* (Freeman, San Francisco), 304.

King, R. H. (1967), The concept of general systems theory as applied to geomorphology; *Albertan Geographer*, **3**, 29–34.

Mann, C. J. (1970), Randomness in nature; *Bulletin of the Geological Society of America*, **81**, 95–104.

McLoughlin, J. B., and Webster, J. N. (1970), Cybernetic and general-system approaches to urban and regional research: A review of the literature; *Environment and Planning*, **2**, 369–408.

Melton, M. A. (1958), Correlation structure of morphometric properties of drainage systems and their controlling agents; *Journal of Geology*, **66**, 442–460.

Ollier, C. D. (1968), Open systems and dynamic equilibrium in geomorphology; *Australian Geographical Studies*, **6**, 167–170.

Rowe, J. S. (1961), The level of integration concept and ecology; *Ecology*, **42**, 420–427.

Simon, H. A. (1965), The architecture of complexity; *General Systems Yearbook*, **10**, 63–76.

Smalley, I. J. and Vita-Finzi, C. (1969), The concept of 'system' in the earth sciences, particularly geomorphology; *Bulletin of the Geological Society of America*, **80**, 1591–1594.

Von Bertalanffy, L. (1956), General system theory; *General Systems Yearbook*, **1**, 1–10.

Von Bertalanffy, L. (1962), General system theory—A critical review; *General Systems Yearbook*, **1**, 1–20.

Wilbanks, T. J. and Symanski, R. (1968), What is systems analysis?; *Professional Geographer*, **22**, 81–85.

2: *Morphological Systems*

"It isn't size that counts so much as the way things are arranged"
E. M. FORSTER: *Howards End*

2.1 The Definition of Morphological Systems

A morphological system is made up solely of the associated physical properties of phenomena: that is to say, their geometry, composition, strength and so forth. In structural terms, morphological systems are the least complex of natural structures. They are, as a result, identifiable in infinitely large numbers and at all scales of magnitude.

Functionally, morphological systems may be isolated, closed or open. Those of interest to the physical geographer are generally closed or open, with the result that many of their properties can be viewed as *responses*, or adjustments to the flows of energy or mass through the cascading systems to which they are linked. This linkage or *interlocking* of morphological and cascading systems will be examined more fully in Chapter 4, but it should be noted that it occurs via physical properties of the former which are either closely correlated with some variable in the cascade or are actually an integral part of the latter; being either storages or regulators within it. The correlation of physical attributes within a morphological system can also frequently be viewed as a reflection of energy relationships within a cascading system.

In general, then, it is not difficult to identify morphological systems within the real world. The problems which arise are:

1 Where to draw the boundaries of each system.
2 How to decide *which* physical properties are of significance:
   (*a*) In terms of the system itself.
   (*b*) In terms of its links (if any) with cascading systems.
3 How to describe the structure of the system.

The answers to the first two questions are largely a matter of common sense, or experience, although there are often serious difficulties in measuring certain physical properties in a meaningful way. Before we consider these questions, it is valuable to look first at the methods which are currently available for the description of morphological systems.

23

2.2 The Description of Morphological Systems

The most valuable tools for the description of morphological systems are those which derive from mathematical *regression and correlation* techniques. The particular form of these techniques which should be used in the analysis of a particular morphological system is dependent, first, upon the questions which the investigator asks about the structure of the system and, second, upon its magnitude.

A. SIMPLE REGRESSION AND CORRELATION ANALYSIS

Causality

The raw data for these analyses consist of *paired* observations of two variables within the system. One is designated the *dependent variable* (Y), and this is thought to be causally related to the *independent variable* (X). This relationship can be most easily visualised in terms of a *scatter plot* of the paired values, X_i, Y_i (Fig. 2.1).

By designating Y, the dependent, and X, the independent, variable, we imply that a change in X always produces a change in Y (but not necessarily *vice versa*) and, further, that values of Y can be *predicted* from values of X; as, for example, the number of cricket chirps per minute can be quite closely predicted from the temperature (Fig. 1.5). The meaning of the covariation of two variables, in terms of cause and effect, can be interpreted in several ways:

(*a*) That X is the cause of Y (Fig. 2.1*a*). This conclusion can usually only be reached on the basis of a genuine understanding of the nature of the processes involved.

(*b*) That there is *autocorrelation* (Fig. 2.1*b*), in other words, that the two variables change in harmony, but without any clear cause and effect relationship between them. This often occurs when each of them is, in fact, the effect of some third variable (Z) which, although a legitimate cause of X and Y, does not appear in the analysis. For example, stream width and meander wavelength are closely correlated, but neither is really the cause of the other, since both are the result of other features of the stream system, most notably bankfull discharge.

(*c*) That X is a cause of Y, but only indirectly, through the agency of one or more variables which do not appear in the analysis.

(*d*) That the relationship between X and Y has appeared only *by chance*, as the result of some partiality in the collection of data. If this is the case, then the collection of a new sample or the enlargement of the sample of paired values of X and Y, should produce a quite different impression of the covariance between them.

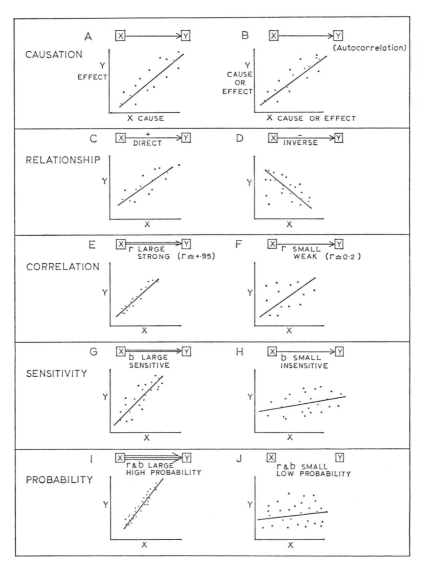

Fig. 2.1   Illustration by scatter plots, linear regressions and symbolic diagrams of the concepts of causation, relationship, correlation, sensitivity and probability between two variables

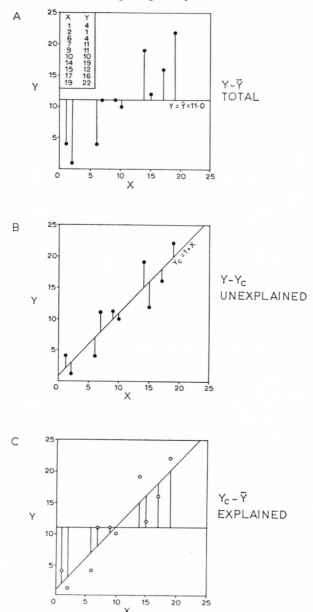

Fig. 2.2   Graphic representation of (*a*) the total sum of squares of the dependent variable (*Y*), (*b*) the deviations (residuals) of the dependent variable from the best-fit linear regression, and (*c*) the extent to which the independent variable (*X*) reduces the total variance exhibited by the dependent variable. (After Krumbein, 1959)

Of course, the concept of 'cause and effect' is very difficult to substantiate, except for a completely isolated system containing only two variables. This never occurs in nature. When we speak of cause and effect, we are simply implying, very often, that Y changes in a predictable manner as X changes.

Regression

The scatter diagram may be viewed as a simple linear *model*. This model may be approximated by fitting a line (a *linear regression*) to the scatter of points, to give the best-fit relationship between X and Y. The best-fit linear regression is of the form:

$$Y_c = a + bX$$

and is calculated so that the sum of squares of all departures of Y from the regression, will be as small as possible (Fig. 2.2). There are several *assumptions* underlying the calculation of the best-fit regression and one of the most important is that the values of Y which correspond to *one* value of X are *normally distributed* in their scatter about the regression line: i.e. the *mean* of their distribution coincides with the position of the regression line (Fig. 2.3).

An absolute measure of the scatter is the *standard error* (σ_{Ys}):

$$\sigma_{Ys} = \sqrt{\left[\sum_{i=1}^{N} (Y_i - Y_{ci})^2 / N \right]}$$

which predicts that, if one could obtain an infinitely large sample of paired values of X and Y, some 68% of these could be expected to fall within + or − $1\sigma_{Ys}$ of the best-fit regression, 95% within + or − $2\sigma_{Ys}$ and 99% within + or − $3\sigma_{Ys}$.

In the best-fit linear regression, 'a' is the calculated value of Y (i.e. = Y_c) when X is zero, and 'b' is the slope of the regression line (Fig. 2.4). The sign of b, the regression coefficient, indicates whether the estimated relationship between X and Y is *direct* or *positive* (an increase in X is associated with an increase in Y, Fig. 2.1c), or *inverse* or *negative* (Fig. 2.1d).

Some relationships are best expressed by semi-logarithmic (Fig. 2.5a), logarithmic (Fig. 2.5b) or polynomial (Fig. 2.6) curves, having the respective forms:

$$\log Y = \log a + \log b.X$$
$$\log Y = \log a + b.\log X$$
$$Y = a + bX + cX^2 + dX^3 \ldots + nX^{(n-1)}$$

A.

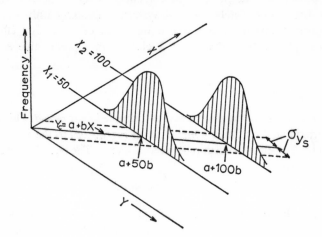

B.

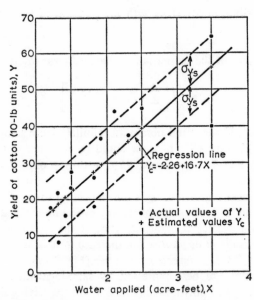

Fig. 2.3   The scatter of points about the linear regression:
(a) Hypothetical frequency curves showing in three-dimensions the relations expressed by the standard error (σ_{Ys}). (After Krumbein and Graybill, 1964)
(b) The standard error associated with a best-fit linear regression line fitted to the yield of cotton (Y) related to the amount of irrigation water applied (X). (After Ezekiel and Fox, 1959)

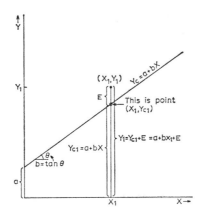

Fig. 2.4   The parameters of the linear regression

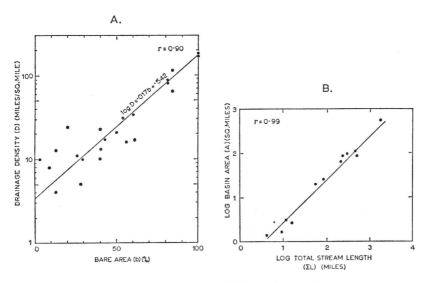

Fig. 2.5   Non-linear regressions (showing coefficients of correlation):
(a) Drainage density as a semi-logarithmic function of percentage bare surface area in a number of drainage basins in the western United States. (After Melton, 1957)
(b) Drainage basin area as a logarithmic function of total stream channel length. (After Morisawa, 1959)

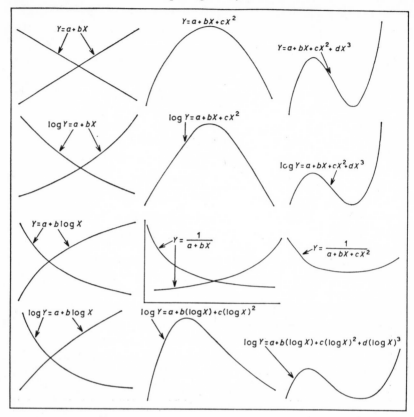

Fig. 2.6   Curves illustrating a number of different types of mathematical functions. First-, second- and third-order polynomial curves are depicted in the top row. (After Ezekiel and Fox, 1959)

Correlation

Most morphological systems contain several variables, interlinked more or less closely. Further, the extent to which a particular X 'explains' a particular Y may vary considerably between one system and another. It is therefore most useful to express the scatter of points about the best-fit regression line not in the units of Y (as is the case if we use σ_{Ys}), but in a dimensionless manner which allows us to compare the degree of association between different variables in one system, or the same variables in different systems. The *coefficient of correlation* (r) is such a dimensionless measure. Values of r vary between ± 1.0 (indicating that all points sampled coincide exactly with the fitted regression line, and the sign of r signifying the sign of b in the regression equation) and 0.0 (indicating that there is a completely random scatter of points and, hence, that there is no correlation between X and Y so that any regression line would provide as good (or bad!) a description of the relationship).

This coefficient of correlation is calculated in terms of the ratio of the *explained variance* of the sample points with reference to the regression line, to the *total variance* (Fig. 2.2*c* and *a*):

$$
r = \sqrt{\dfrac{\displaystyle\sum_{i=1}^{N} [(Y_i - Y_{ci})^2]/(N-1)}{\displaystyle\sum_{i=1}^{N} [(Y_i - \bar{Y})^2]/(N-1)}}
$$

and may be viewed as a measure of how much better the best-fit regression line describes the plot of points than does a line of zero slope (i.e. $b = 0$) passing through the mean value of Y (i.e. $a = \bar{Y}$). The square of the coefficient of correlation gives the *coefficient of determination* (r^2) which expresses, as a percentage, the amount of variance in Y 'explained' by the observed variance in X. The *unexplained variance* is therefore $(1 - r^2) \times 100\%$ (see Fig. 2.2*b*).

This way of analysing the scatter of points about a regression line has the further implication that the best-fit regression of Y upon X (i.e. $Y_c = a_{yx} + b_{yx} X$) is different from that of X upon Y (i.e. $X_c = a_{xy} + b_{xy} Y$) (Fig. 2.7). However, although two different regression equations can be calculated, the value of r obtained for both is the same and its magnitude is a measure of how closely the two regressions coincide. This is because $r = \sqrt{b_{yx}.b_{xy}}$ and both regressions tend towards identical forms ($Y = a + bX$ and $X = -a/b + Y/b$) as r approaches unity.

Noise

The best-fit regression $Y_c = a + bX$ gives only an idealised and generalised expression of the relationship between X and Y, which may be more accurately stated as $Y_c = a + bX + E$, where E is an error term, expressing the deviation of each point from the regression line. It can be seen that E (see Fig. 2.4) is another way of describing the unexplained variance (Fig. 2.2*b*) and it is often referred to as *noise* (by analogy with radio static which is superimposed on, and confuses, the information signal). Noise in regression data can come from a number of sources (Fig. 2.8) and may therefore contain different components:

1 Standard error noise (see Fig. 2.3) arising solely from the assumed normal distribution of the scatter of Y about $Y_c = a + bX$. There are many possible reasons for this, varying from errors in measurement to some assumed natural variation inherent in the phenomena being expressed.

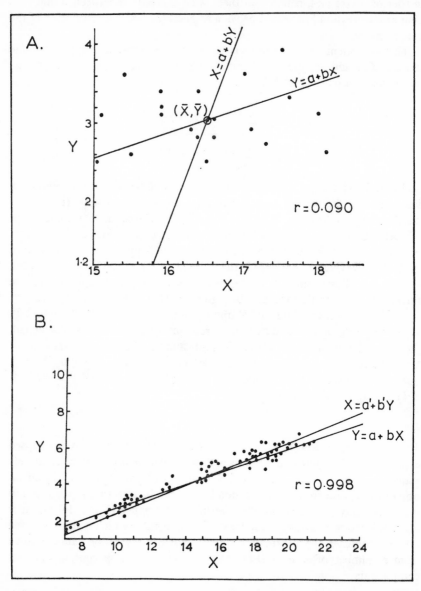

Fig. 2.7   Calculation of best-fit linear regressions of *Y* as a function of (i.e. upon) *X*, and of *X* as a function of *Y*. (*a*) shows this in respect of a scatter of points exhibiting low correlation ($r = 0.090$) and (*b*) in respect of a scatter exhibiting high correlation ($r = 0.998$). (After Olson and Miller, 1958)

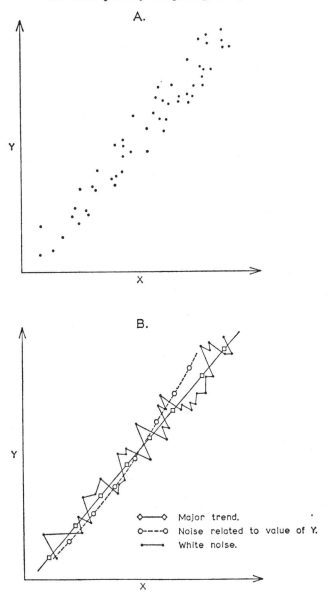

Fig. 2.8   An attempt to illustrate how a scatter of points (*a*) can (*b*) embody simultaneously information regarding a major trend (i.e. a relationship between *Y* and *X*), noise relating to the value of *Y*, and a random 'white' noise. Of course, *each point* does not separately depict one of these three types of information and the representation in (*b*) is schematic only

2 Other-variable noise. This noise varies in amount with the particular value of Y which is considered, suggesting that one or more other independent variables (X_2, X_3 ... etc.) are exerting an influence on Y. This matter will be considered in more detail later in the chapter.

3 White noise. Many data sets show a random, featureless noise, the values of which are independent of Y or X. It is difficult to decide how far white noise may be related to sources 1 and 2 (or to the combined effect of a large number of independent variables which may, individually, be unimportant) and many would consider that this is evidence of the obscure operation of variables other than those included in the existing analysis. It is a common feature of causal investigations that today's noise is tomorrow's explained variance!

It may be, of course, that part of the random variation in Y is the result of random fluctuations in one or more of the controlling variables. However, it is by no means certain that such random fluctuations in an independent variable should necessarily produce corresponding variation in Y, since they may be too small, of too short a duration, or below the minimum threshold to produce an effect.

Probability

The coefficient of correlation (r) gives a measure of how well the best-fit regression line fits the data (Fig. 2.1 e and f) but it is the regression coefficient (b) which expresses the nature and sensitivity of the relationship (Fig. 2.1 g and h), as the larger the value of b, the greater the change in Y which is expected to follow a change in X.

It must be remembered that the scatter of points on the graph merely represents, in most cases, a *sample* of a very large (probably infinite) population of possible points. Since this is so, it is most important to determine the probability that r differs significantly from zero: in other words, the probability that repeated sampling of N points at random from the population will continue to produce as great a value of r as that observed in the sample. To do this we assume that the total population of paired values of X and Y is really random with $r = 0$, and that an observed value of r of other than 0 is simply due to the vicissitudes of chance sampling. The probability (p) that such a non-zero value of r could have been obtained by randomly drawing a sample of N variates from a completely randomly-distributed population may be estimated by applying the statistic:

$$t_{(N-2 \text{ degrees of freedom})} = r\sqrt{(N-2)}/\sqrt{(1-r^2)}$$

(The degrees of freedom here equal $N - 2$ because there are two constants in the best-fit equation: a and b.)

Where p falls as low as 5%, it is considered that the observed value of r is too large to have been obtained merely by chance and, consequently, that it represents a real correlation between X and Y. Because of the close relationship between r and b, a value of r which differs significantly from zero implies that b is significantly different also (Fig. 2.1*i*). Values of p of ·05 (5%), ·01 (1%) and ·001 (0.1%) may be indicated on a correlation diagram by links composed of one, two and three lines, respectively.

B. CORRELATION SYSTEMS

The methods of simple regression and correlation analysis and the interpretation of their results, illustrated by the conventions shown in Fig. 2.1, can be used to construct *correlation systems*, developed from groups of variables which it is thought may possibly intersect. Tests of intersection produce systems in which each variable is related to a number of others: at stated levels of r and p; by one type of relationship ($+$ or $-$); and in a suggested direction of causality. It is a further characteristic of such correlation systems that one or more of the component variables is correlated with variables outside the system.

The simplest type of correlation system is that in which one can assume *additivity*: that is, each controlling variable exerts a distinct and separate effect on those variables dependent upon it. One such simple real-world system is shown in Fig. 2.9*a* and the matrix of r^2 values is set out in Fig. 2.9*b*. Although variables X_2, X_3 and X_4 exhibit additivity (i.e. their total r^2 values sum to less than unity), the relationships involving X_1 and X_5 appear slightly non-additive (i.e. $\Sigma r^2 > 1$). Such departures from true additivity are rather common in natural systems and, in fact, that represented in Fig. 2.9 comes unusually close to the ideal. It is possible to approach closer to a true additive system from the simple correlation matrix (Fig. 2.9*a*) by discarding progressively those links of lowest correlation, by inserting directions of causality on the basis of rational theory, testing the actual versus the expected values of r for all non-linkages, making changes where the greatest discrepancies occur between actual and expected values, retesting the new model and so on, until a satisfactory accommodation has been arrived at (Fig. 2.9 *c–e*).

Most correlation systems in physical geography are non-additive, in that the combined effect of the causative factors is, commonly, less than the sum of the individual effects (i.e. there is *replication*) and, occasionally, greater (i.e. there is *reinforcement*). Where non-additivity occurs, the simple construction of correlation systems described above has to be replaced by more sophisticated analysis. Firstly, only those links which represent values of $p \leq$ ·05 are retained in the correlation system and, secondly, no directions of causation are shown, since with purely geometrical variables which constrain each other it is difficult to say which causes which. By and large, subsequent

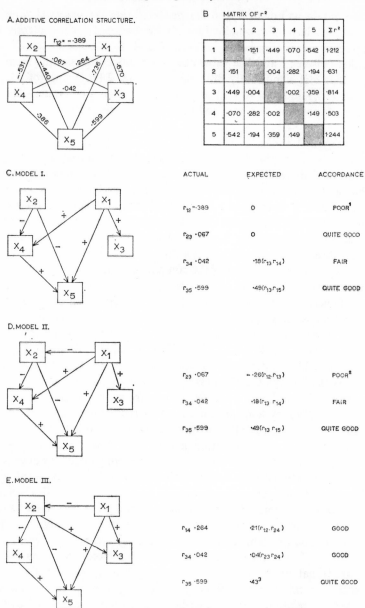

Fig. 2.9   An example of a more-or-less additive correlation matrix (*b*) and its correlation structure (*a*), illustrating the progressive comparison of reality against models of the way in which the system should interlock (*c, d, e*). (After Blalock, 1964)

analysis proceeds intuitively, with limitations set by the investigator's degree of insight into the manner in which the particular system functions. However, in some cases, it is possible to take the statistical analysis further by means of a *cluster analysis* of the correlation matrix. This may produce pairs of variables which are more highly correlated with each other than with any other variable, each pair proving the focal point of a cluster of other factors, the clusters themselves being frequently connected by links between 'outer' variables.

By and large, the analysis of non-additive correlation structures is the most basic and difficult procedure in the analysis of morphological systems in physical geography and we shall return to the discussion of this question later in the chapter.

C. MULTIPLE CORRELATION

Often the very nature of an investigation of the physical world prompts one to identify a system in which one of the variables occupies a highly nodal position in terms of direction of causation, such that it is important to see how far changes in this dependent variable are conditioned by interactions of the independent variables. In such a case we must ask: How many major independent variables are involved in accounting for variations in the dependent variable? Are the independent variables interrelated and how do their relevant interactions operate? What is their relative importance? Does the importance of a given variable vary from situation to situation? All these questions may be answered by analyses of multiple linear correlation.

Just as it is possible to fit a line in two dimensions of the form $Y_c = a + bX$ to a scatter of points on a graph, it is equally possible to calculate a best-fit plane of the form: $Y_{c.12} = a_{Y1.2} X_1 + b_{Y2.1} X_2$ to a three-dimensional scatter of points (Fig. 2.16), by solving the three simultaneous equations:

$$\Sigma Y = N a_{Y.12} + b_{Y1.2} \Sigma X_1 + b_{Y2.1} \Sigma X_2$$

$$\Sigma YX_1 = a_{Y.12} \Sigma X_1 + b_{Y1.2} \Sigma X_1{}^2 + b_{Y2.1} \Sigma X_1 X_2$$

$$\Sigma YX_2 = a_{Y.12} \Sigma X_2 + b_{Y1.2} \Sigma X_1 X_2 + b_{Y2.1} \Sigma X_2{}^2$$

The calculation of total, explained and unexplained variance by the use of the values, Y, Y_c and $\bar{Y}$ then allows the calculation of the *coefficient of multiple correlation* ($R_{Y.12}$) and, just as in the 2-dimensional case, the *coefficient of multiple determination* ($R_{Y.12}^2$) gives the percentage of variation in Y which is accounted for, or 'explained', by the combination of the variable X_1 and X_2. In this instance

$$R_{Y.12}^2 = [\Sigma Y_{c.12}^2 - \bar{Y}\Sigma Y]/[\Sigma Y^2 - \bar{Y}\Sigma Y] =$$

$$[a_{Y.12}\Sigma Y + b_{Y1.2}\Sigma YX_1 + b_{Y2.1}\Sigma YX_2 - (\Sigma Y)^2/N]/[\Sigma Y^2 - (\Sigma Y)^2/N]$$

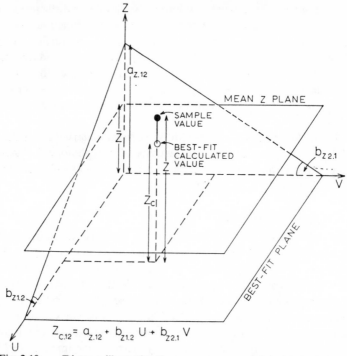

Fig. 2.10   Diagram illustrating the parameters associated with the fitting of a plane to a three-dimensional scatter of points related to U, V, Z axes. Note the similarities with the two-dimensional cases depicted in Figs. 2.2 and 2.4 (In trend-surface analysis the axes U, V, Z are substituted for X_1, X_2, Y)

Although it is impossible to depict or visualise higher order multiple correlations, it is equally possible to calculate theoretical 'linear surfaces' relating one dependent to three or more independent variables and, similarly, to calculate values of $R^2_{Y.123\ldots n}$ giving the percentage of explanation of Y in terms of all combinations of $X_1, X_2 \ldots X_n$. This shows immediately the effects of individual independent variables upon Y, how they vary when combined with others, and how complete an explanation of the variation of Y is given by all the factors assumed to form part of its explanatory system.

It is a feature of multiple-correlation systems in physical geography that there is considerable replication between the independent variables. This creates problems, for it is possible by chance to omit from the analysis that variable which is genuinely 'causing' the observed variation in Y, but to obtain a satisfactory explanation of that variation in terms of other 'independent' factors which are similarly under the control of the excluded variable.

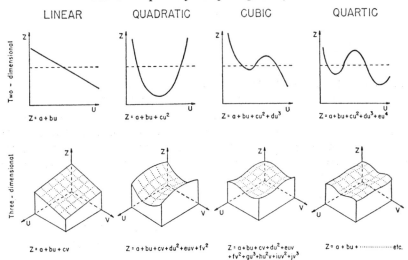

Fig. 2.11   The relationship of four orders of two-dimensional polynomial curves to their three-dimensional counterparts. (From Chorley and Haggett, 1965)

It is most important to bear this fact in mind when carrying out multiple regression analyses, or when considering the results of such investigations.

D. SPACE FILTERING

An interesting extension of the above multi-variate analysis occurs when one is dealing with a spatial system in which the two independent variables (U, V) are locational co-ordinates on the ground and the dependent variable (Z) represents some property which varies across the surface. Fig. 2.10 illustrates a simple linear surface fitted in this way. In more complex cases 3-dimensional polynomial *trend surfaces* can be fitted to the spatial data (Fig. 2.11) and it is possible to identify large-scale regional variations and associate their patterns with similar spatial variations in possible causal factors. (One might find, for example, that irrespective of rainfall, drainage density varies inversely with mean grain size of the underlying sandstone in a region.) Such a fitted polynomial surface (for which a value of R^2 can be calculated) thus expresses some regional response to some cause which is assumed to operate spatially. Further, the regional surface may be used as a *filter*, by plotting the *deviations* of Z from Z_c (i.e. removing the regional trend). This will produce a map of local *residuals* (i.e. *spatial noise*), the distribution of which may enable one to infer the operation of additional local processes which are otherwise masked by the regional trend. This topic is discussed more fully in Chapter 4.

E. ANALYSIS OF VARIANCE

In few, if any, natural morphological systems is it possible to predict the value of Y for particular value(s) of the independent variable(s) exactly: that is to say, there is always a component of 'noise'. We measure the amount of noise as: $1 - r^2$ or $1 - R^2$. Although we have already outlined the principal sources of unexplained variation in Y, there is a further systematic source which we have not yet considered. This is noise which arises from the inclusion—within one sample—of systems operations on either side of a critical threshold: that is, separated by a discontinuity in the *magnitude* of the system.

An illustration of the effects of thresholds comes from the relationship between channel depth (X) and channel width (Y); considered (a) at one station through time, or (b) at one time along a river.

(a) At some stage, known as bankfull discharge, the river will exactly fill its bed; below this discharge depth will change more rapidly than width; above this discharge overbank flooding will occur and width will change more rapidly than depth (Fig. 2.12). The sensitivity of the association of X and Y will therefore vary substantially in conditions of between- and over-bank flow.

(b) As we proceed down-river there is likely to come a point where the river ceases to flow in bedrock and a flood-plain develops. Upstream of this point the channel will be relatively deep and narrow, so that width will be comparatively insensitive to changes in depth; downstream of this point, width will vary more radically for a given change in depth. The sensitivity of the relationship between X and Y will therefore vary substantially with the nature of the material forming the channel.

In both cases, the threshold within the system is rather easy to identify.

We would tend, therefore, to consider that there are discrete morphological systems at below and above bankfull discharge; or pertaining to bedrock and alluvial channels. If we identify these subsystems and the thresholds between them rather precisely, we would expect to find a close relationship between X and Y within each subsystem (i.e. $1 - r^2$ is low). However, if we were unaware of the existence of thresholds within the system and collected paired observations of depth and width irrespective of discharge or position along the stream, the best-fit regression line would average the trends in each

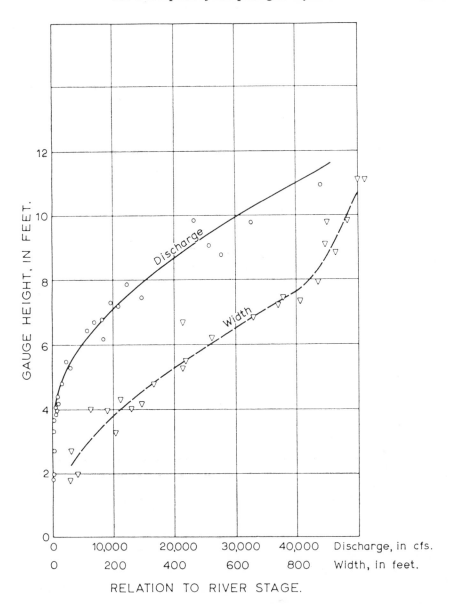

Fig. 2.12   An illustration of the passage of a threshold in a system and its effect upon the relationship of two variables. As discharge increases to bankfull and beyond, the relationship between channel width and depth changes dramatically. (After Leopold and Maddock, 1953)

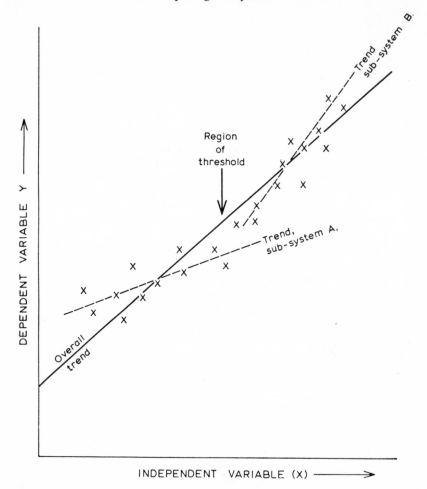

Fig. 2.13   Diagrammatic illustration of the way in which two different trends and the threshold between them may be masked by the calculation of an average trend

subsystem, with the result that $1 - r^2$ would be rather large (Fig. 2.13). It is unfortunately true that, in many cases, we have little or no prior knowledge of the critical thresholds within morphological systems and, consequently, may accept low values of r^2 or R^2 as the result of purely random variations. How should we deal with this problem?

No Prior Knowledge of Thresholds

In this case, the only possible strategy is to collect our observations in very large numbers and to plot the scatter of points for every pair of variables.

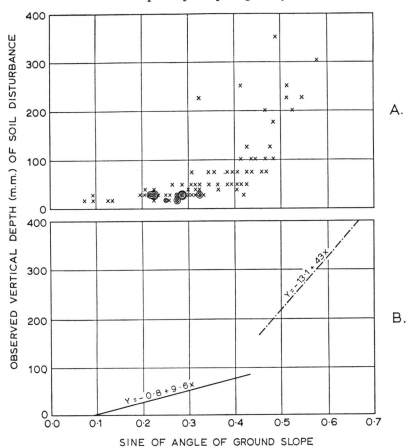

Fig. 2.14   Example of a threshold:
(*a*) Shows the plot of depth of soil disturbance (*Y*) against sine of angle of ground slope (*X*)
(*b*) Illustrates the two, quite distinct, relationships between *X* and *Y* on either side of the threshold region where *X* = 0·45. (From Kennedy, 1969)

This may be a time-consuming process, but it should reveal discontinuities within the system. Fig. 2.14 shows one such discontinuity, revealed in this way.

Some Prior Knowledge of Possible Thresholds

In certain cases, although we cannot be sure that thresholds exist, we may still have an intuitive impression of the positions in which they are most likely to occur. That is, we may feel that changes in the magnitude of particular variables may be critical for the sensitivity of their relationship with others;

5

or that qualitative changes in the location of the system in space may similarly represent the passage from one subsystem to another.

Where such hypotheses can be formed, it is possible to test the uniformity of a system, before carrying out correlation analysis, by means of *analysis of variance* techniques.

Let us assume that we are interested in establishing the form of the relationship between channel depth (X) and channel width (Y) in a group of rivers of similar size in one area, but flowing on k different lithologies. It is possible that the nature of the bedrock could influence the channel form, through the size of debris. If bedrock type has no effect, we would expect to find the same average width/depth ratio on all rock types and approximately the same degree of dispersion about the average value. If bedrock has a systematic effect, however, either the mean ($\bar{X}$) width/depth ratio or the variance (σ^2) (the average of the squared deviations ($X - \bar{X}$) from the mean), or both, will differ significantly between lithologies.

To test the uniformity of our system, we first take a sample of width/depth ratios from rivers on each lithology: it is generally helpful to take samples of equal size (m). Secondly, we calculate the mean ratio for each lithology and the variance of each sample. The next step is to compare the k sample variances: the simplest test of this is *Cochran's Test:*

$$C_{(k\&m-1)} = \sigma^2(\text{largest})/\left[\sum_{j=1}^{k} \left(\sigma^2_{j=1} + \sigma^2_{j=2} + \ldots + \sigma^2_{j=k} \right) \right]$$

(C has k and $m - 1$ degrees of freedom, where m is the size of each sample: this test requires equality of sample size.)

If the k sample variances do not differ significantly, we may proceed to test the hypothesis that the k sample means are statistically similar. We do this by a *one-way analysis of variance*, where the null hypothesis is:

$$H_0 : \bar{X}_1 = \bar{X}_2 = \bar{X}_3 \ldots = \bar{X}_k$$

By pooling the k samples to give one sample of size N, we can calculate an overall sample mean ($\bar{X}$) and an overall sample variance

$$\hat{\sigma}^2 = \sum_{i=1}^{km} [(X_i - \bar{\bar{X}})^2]/(km - 1)$$

If our null hypothesis is correct then $\hat{\sigma}^2$ is exactly equal to the variance within each sample, or,

$$\hat{\sigma}^2 = \hat{\sigma}_m^2 = \sum_{i=1}^{k} \left[\sum_{i=1}^{m} (X_i - \bar{X}_j)^2 \right]/(N - k)$$

(see Fig. 2.15*a*). If our null hypothesis is not correct, i.e. $\bar{X}_1 \neq \bar{X}_2 \neq \bar{X}_3 \neq \ldots \bar{X}_{k2}$, then $\hat{\sigma}^2$ will be *larger* then $\hat{\sigma}_m^2$ (Fig. 2.15*b*): the difference $\hat{\sigma}^2 - \hat{\sigma}_m^2$ is ascribable to the fact that $\bar{X}_k \neq \bar{X}$, or to the variance of the sample means themselves

$$\hat{\sigma}_k{}^2 = \sum_{j=1}^{k} [(\bar{X}_j - \bar{\bar{X}})^2]/(k-1)$$

Just as in correlation analysis we determine the significance of the 'control' of Y by X by testing the ratio explained/unexplained variance; so the

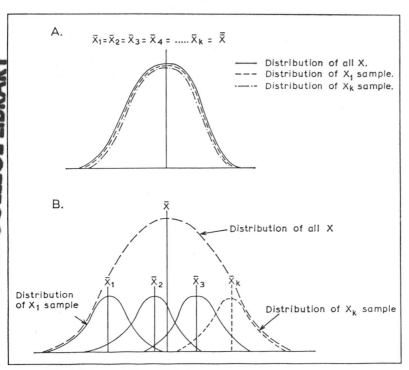

A.

$\bar{X}_1 = \bar{X}_2 = \bar{X}_3 = \bar{X}_4 = \ldots \ldots \bar{X}_k = \bar{\bar{X}}$

—— Distribution of all X.
– – – Distribution of X_1 sample.
–·—·· Distribution of X_k sample.

B.

$\bar{\bar{X}}$

—— Distribution of all X

$\bar{X}_1 \quad \bar{X}_2 \quad \bar{X}_3 \quad \bar{X}_k$

Distribution of X_1 sample

Distribution of X_k sample

Fig. 2.15   The relationship between the difference of sample means ($\bar{X}$) and the variability of the distribution of all samples, considered together.
(*a*) Illustrates the case where all samples have equal means and equal variances.
(*b*) Demonstrates the case where, although sample variances remain equal, sample means differ.

analysis of variance establishes the significance of the 'control' of X by some exterior factor, by testing the ratio $\sigma_m{}^2/\sigma_k{}^2$, or 'Between Sample'/'Within Sample' variance:

$$F_{(k-1;\,n-k)} = \frac{\displaystyle\sum_{j=1}^{k} [\bar{X}_j - \bar{\bar{X}}]/(k-1)}{\displaystyle\sum_{j=1}^{k} \left[\sum_{i=1}^{m} (X_i - \bar{X}_j)^2 \right]/(N-k)} = \frac{\hat{\sigma}_k{}^2}{\hat{\sigma}_m{}^2}$$

If the *F* ratio proves significant, we must accept the fact that the control—in the example given, variable lithology—has some systematic influence upon the average values of *X* (channel width/depth ratio). In this case it is entirely possible that we are dealing with morphologically different subsystems and it is safest to proceed on this assumption.

There are many other forms of analysis of variance—*nested, factorial, Latin square* models and so forth—which will evaluate systematic variations in one variable under the joint influence of several external controls. These models obviously allow a much more precise location of different subsystems than the one-way model, but demand more accurate knowledge of the probable thresholds within the major system so that it is not always possible to use them. (The use of factorial designs is described in Chapter 4.)

Whether one can or cannot employ analysis of variance, it is extremely important to check that morphological systems do not contain critical thresholds, *before* carrying out any form of correlation analysis.

2.3 The Identification of Morphological Systems

We have dealt, at length, with the procedures which allow us to describe the structure of morphological systems, but we have not yet considered the way we would set about identifying such structures in the landscape, nor described the forms in which they might appear. The concluding sections of this chapter attempt an outline of both topics.

In order to examine the variety of systems that can be identified, let us consider an idealised section of the landscape (Fig. 2.16): a series of parallel river valleys draining to a cliffed sea-coast. (We shall not specify the scale of the landscape section, except in purely relative terms—to state that all the major valleys contain at least 4th order streams—but it should be borne in mind that the difficulties of carrying out the investigations described below will be very different if the area is 1000 sq. miles than if it is 1000 sq. feet.) What are the major morphological systems we might identify within this landscape?

A. DRAINAGE BASIN MORPHOMETRY

A drainage basin is clearly an organized section of the land surface and, as such, its geometric features are very likely to be functionally related. However, we cannot establish the nature of morphological relationships from looking at *one* drainage basin, but need to take a sample of several. At this point we should recall that the order of the drainage basin (see Fig. 1.1b) is very likely to have some effect upon morphometry, as it will relate in some manner to the magnitude of discharge. We should, then, consider that the

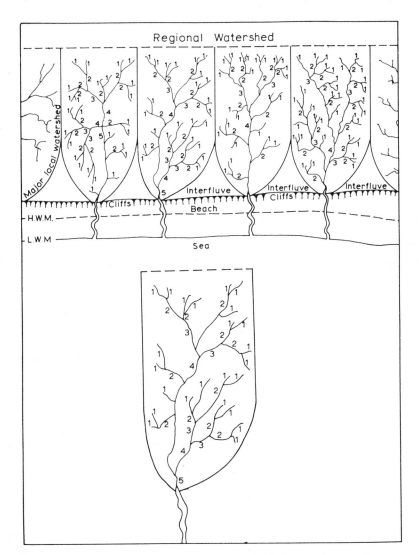

Fig. 2.16 An hypothetical landscape, showing simple 4th and 5th order streams draining directly to the sea coast. Inset is an enlargement of one 5th order basin, with its drainage channels ordered according to the Strahler system

4 large drainage basins shown in Fig. 2.16, provide us with the following samples:

$$27 + 24 + 24 + 34 - \text{1st order basins}$$
$$11 + 11 + 10 + 14 - \text{2nd order basins}$$
$$4 + 4 + 2 + 4 - \text{3rd order basins}$$
$$2 + 2 + 1 + 1 - \text{4th order basins}$$
$$1 + 1 - \text{5th order basins}$$

These represent five possible subsystems of drainage basins. Which of their geometric properties might be of interest?

1 The perimeter (watershed) length (p), in miles.
2 The area (a), in square miles.
3 The maximum elevation within the basin (H_{max}), in feet.
4 The minimum elevation within the basin (z), in feet.
5 The relief of the basin ($H_{max} - z = r$), in feet.
6 The relative relief of the basin ($100r/p5280 = R$), in feet.
7 The circularity of the basin (a/area of circle with same $p = C$).
8 The total length of stream channels of order u (l_u), in miles.
9 The total length of all stream channels (Σl_u), in miles.
10 The drainage density ($\Sigma l_u/a = D$), in miles per square mile.
11 The number of stream channels of order u (n_u).
12 The total number of all stream channels (Σn_u).
13 The stream frequency ($\Sigma n_u/a = F$), as number per square mile (Fig. 2.17).
14 The ruggedness number ($D.r/5280 = H$).
15 The bifurcation ratio ($n_u/n_{u+1} = R_b$).
16 The hypsometric integral (I) = $\int_0^{100} \dfrac{da}{dr}$ (Fig. 2.18).

All these quantities are, obviously, precisely fixed for each basin at any one time (or mapped on any particular scale) but it has been found that all are rather difficult to measure accurately, whether in the field or from maps. The most difficult values to fix exactly are 2, 9, and hence 10. In addition to these 16 geometric properties of drainage basins, there are two further characteristics which we might consider of interest: the maximum angle of valley-side slope (θ) and the channel gradient (γ). However, these two last features obviously cannot apply realistically to the whole of each basin, but must be derived as the average of samples of some particular size, arranged at random within each basin: they are, then, of a different order of accuracy than variables 1 to 16.

These 16 properties are by no means the only variables in the basin morphometry system—one can, for example, create more indices like D by combining two or more other variables, and these indices then become

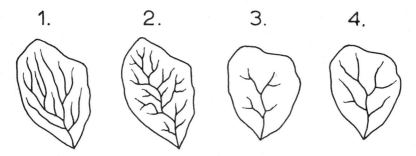

Fig. 2.17   Basins (1) and (2) have the same drainage density, but different stream frequencies; whereas basins (3) and (4) have the same stream frequency but different drainage densities. (From Strahler, 1964)

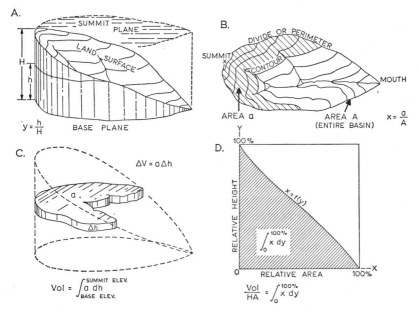

Fig. 2.18   Diagrams illustrating the calculation of the hypsometric integral. (From Strahler, 1952)

(*a*) The derivation of the height values
(*b*) The derivation of the area values
(*c*) An integral slice of drainage basin volume
(*d*) The plotted hypsometric curve

morphometric measures. However, between them the listed variables cover: the areal extent and shape of the basin (determining the amount of precipitation and insolation received); the shape and overall slope (determining the rapidity of runoff); and the density of the channel network (determining the efficiency with which runoff is removed and the amount of relatively inefficient unconcentrated surface wash and throughflow). All three sets of features should also reflect the balance between rock type, rock structure and climatic conditions.

B. SLOPE MORPHOMETRY

It seems likely that there will be three distinct sub-systems of slopes within this section of the landscape (Fig. 2.16).

That of the slopes of the interfluve areas.

That of the valley-side slopes.

That of the sea cliffs.

The separation of these sub-systems would seem to be advisable, since the magnitude of the down-slope component of gravity (represented by the sine of slope angle) is likely to vary markedly from one to the other and, as a result, the degree of vegetation cover and the balance between different processes of subaerial denudation is also likely to differ substantially.

Interfluve Areas

The largest continuous interfluve areas in this section of the landscape are those marked as lying immediately behind the cliffed coastline: in other areas such slopes might represent undissected portions of the upland surface towards the head or sides of drainage basins. Whatever the site of such slopes, one initial difficulty lies in accurately delimiting their areal extent: as a working guide we may say that they occupy zones adjacent to local or regional watersheds and separated from the major drainage lines by pronounced convexities in the valley-side slope profiles. The second problem is that we have to sample the characteristics of these slopes in some way. Having defined the position of interfluve areas, we can superimpose a graticule and, by selecting the successive two- or four-digit numbers from random number tables and using these as northings and eastings, we can locate a sample of points within the interfluve zones. From this point, it is almost obligatory to proceed by fieldwork rather than map analysis, although *some* of the characteristics of such slopes may be determined from large-scale aerial photographs.

At each point we need to determine the line of maximum inclination: having done so, it is simplest to measure off a fixed distance from the original point. This will give an arbitrary 'slope', but as gradients in most interfluve areas are very slight this is preferable to searching for 'breaks of slope'.

The variables which are of primary interest in the interfluve-slope morphometry system are:

1 The angle of slope (θ).
2 The depth of soil or weathered mantle (W).
3 The principal grain-size characteristic of the soil or mantle: those sizes than which 84%, 50% and 10% of the material, by weight, is finer (D_{84}; D_{50}; D_{10}).
4 The pH value of the soil or mantle (pH).
5 The moisture content of the soil or mantle, as a per cent of its dry weight (S_M).
6 The porosity or void ratio of the soil or mantle (V_R).
7 The organic content of the soil or mantle, by weight (W_0).
8 The root weight of the soil or mantle (W_R).
9 The degree of surface cover by plants (S_c), in per cent.
10 The average height of the vegetation (H_v).

It is noticeable that, with the exception of surface slope, all these variables must be estimated by averaging conditions along the line of slope in some way. Moreover, the values of several (e.g. 4, 5, 9, 10) will be highly dependent upon the time of year at which the study is carried out: that is to say, these variables represent links between the 'inert' morphological system of the slope and the dynamic cascading systems of solar energy and the hydrologic cycle.

Clearly, despite its apparent simplicity, the interfluve slope subsystem is in reality a great deal more complex and subtle than the drainage basin morphometry system.

Valley-side Slopes

Having defined the extent of the interfluve slope system, we have automatically defined that of the valley-side slopes as well, but several problems remain. First, because streams of differing orders presuppose differing discharges, there is a strong suggestion that the amount of downcutting (or degree of maturity) will also differ with order and hence, that we need to delimit five different subsystems of valley-side slopes, each one corresponding to one of the stream orders. Second, since most valley-side slopes are genetically related to fluvial erosion, it follows that we may expect differences in the morphological systems of slopes which are actively undercut by streams and of those which are basally protected: this distinction in *local erosional environment* will be most pronounced along streams which are meandering and/or have developed floodplains. Finally, there is still one other possible source of systematic noise: the rate of weathering of the bedrock, depth of soil cover, nature and extent of vegetation and, hence, speed of operation of subaerial denudation, will all reflect the magnitude of input from the cascade systems of solar radiation and the hydrological cycle (see Chapter 3). This means that the *orientation* of the slope may prove to exert a critical control over the nature of the morphological system involved. If this is so, we would need to recognise perhaps four further sub-systems of valley-side slopes, corresponding to the four quadrants of the compass.

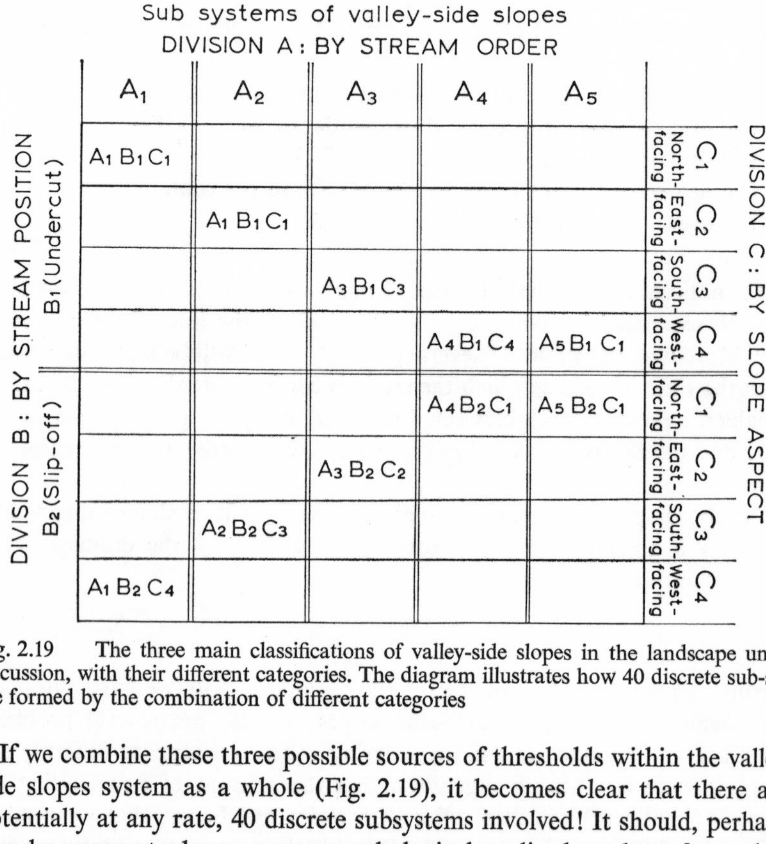

Sub systems of valley-side slopes
DIVISION A: BY STREAM ORDER

	A_1	A_2	A_3	A_4	A_5		
DIVISION B: BY STREAM POSITION — B_1(Undercut)	$A_1 B_1 C_1$					North-facing C_1	**DIVISION C: BY SLOPE ASPECT**
		$A_1 B_1 C_1$				North-East-facing C_2	
			$A_3 B_1 C_3$			South-East-facing C_3	
				$A_4 B_1 C_4$	$A_5 B_1 C_1$	South-West-facing C_4	
B_2(Slip-off)				$A_4 B_2 C_1$	$A_5 B_2 C_1$	North-facing C_1	
			$A_3 B_2 C_2$			North-East-facing C_2	
		$A_2 B_2 C_3$				South-East-facing C_3	
	$A_1 B_2 C_4$					South-West-facing C_4	

Fig. 2.19   The three main classifications of valley-side slopes in the landscape under discussion, with their different categories. The diagram illustrates how 40 discrete sub-sets are formed by the combination of different categories

If we combine these three possible sources of thresholds within the valley-side slopes system as a whole (Fig. 2.19), it becomes clear that there are, potentially at any rate, 40 discrete subsystems involved! It should, perhaps, now be apparent why so many morphological studies have been focused on this particular section of the landscape.

Having identified the possible subsystems we are immediately faced with the practical difficulty of sampling the valley-side slopes within each one. We *could* resort to the same sort of technique which was suggested for the interfluve slopes, but this would be an extremely difficult and time consuming task. A valid and much more feasible alternative is demonstrated in Fig. 2.20.

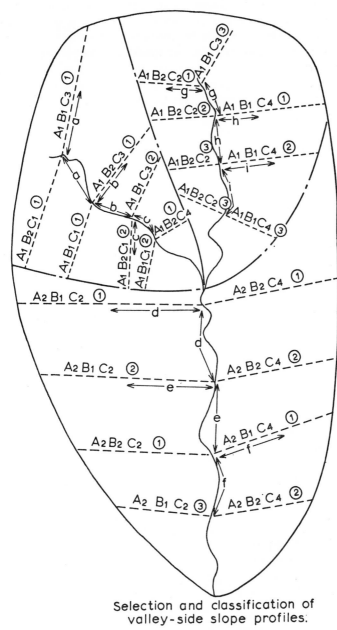

Selection and classification of
valley-side slope profiles.

Fig. 2.20   An illustration of one method of sampling valley-side slope profiles, and classifying them into subsets. See the text for a detailed explanation of the procedure

Within each section of the valley an arbitrary or, better, random starting point (*q*) near the head of the reach is selected and a cross-profile measured. Each *half* of this profile is labelled by: the order of stream; the local erosional environment; orientation: and then numbered. One half of the length of the longer of these profiles is then measured down valley, a new cross-section located and the procedure repeated. When all the basins to be sampled have been treated in this way, samples may be drawn of each sub-system of slopes, by using random number tables.

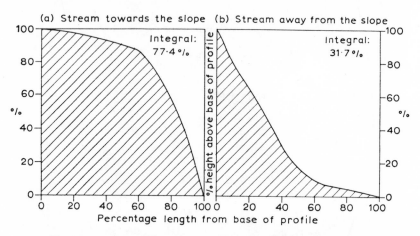

Fig. 2.21   Examples of the height/length integrals of different categories of valley-side slopes in southern Manitoba. (From Kennedy, 1969)

Which variables may be relevant to the morphology of the valley-side slope system?

1 The Strahler maximum angle of slope (θ): that is, the angle of the steepest part of the slope, more than 5 feet in length, which does not represent a cliff of any type.
2 The absolute maximum angle of slope (β).
3 The height of the profile, from channel to interfluve (H_t).
4 The length of the profile (Lg).
5 The average angle of slope (H_t/Lg = sine α).
6 The length of the Strahler maximum angle section (L_{SM}).
7 The height/length integral: $\left(\int_0^{100} dLg/dHt = H.L. \right)$. This assesses the degree convexity or concavity of the profile and is particularly sensitive to changes in the length of the lower concave section (see Fig. 2.21).

8 The gradient of the stream channel at the base of the profile (γ).

9a & 9b. The depth of soil or weathered mantle: (a) on the maximum angle section and (b) over the slope as a whole (W; W').

10a & 10b The D_{10} value of the soil or mantle: ditto.

11a & 11b The pH value: ditto.

12a & 12b The soil moisture value: ditto (S_M; S_M').

13a & 13b The porosity or void ratio: ditto (V_R; V_R').

14a & 14b The organic content: ditto (W_0; W_0').

15a & 15b The root weight: ditto (W_R; W_R').

16a & 16b The percentage of vegetation cover: ditto (S_C; S_C').

17a & 17b The average height of vegetation: ditto (H_v; H_v').

These variables outline a rather complex morphological system, but in fact, we could justifiably add those of the channel geometry (see Chapter 2.3c), since it is a little arbitrary (and quite difficult, as we shall see) to separate slope from stream or to consider only one of the features of the channel morphology system—bed gradient—as 'important' to the form of the slope.

Why is the valley-side slope system so complex? Basically because it represents both a transition in form—between the interfluve and the stream channel—and a transition in function—it is a zone of debris production and also a region across which material is transported—and, in addition, because most valley-side slopes are linked to both an interfluve and a channel system so that their development is conditioned by the speed and direction of change in these adjacent systems (see Fig. 2.22). We might expect that the relative importance of geometric and non-geometric variables within the valley-side system would reflect the varying strength of the tendencies to 'stream-ness' and 'slope-ness', i.e., as transport down the slope to the stream becomes of increasing importance, so do the dimensions of the profile: as the weathering processes dominate, so the transport decreases and the non-geometric features of mantle, soil and vegetation assume greater significance. Thus, if our area of study were to contain dry valleys we might expect that their slopes would belong to morphological systems more akin to that of the interfluve areas, than that of the main valley-sides.

Sea Cliffs

The main distinction between this sub-system and the other two is the differing 'intensity' of slope angles to be expected and, therefore, the greater the probable importance of bedrock type and structure and the less the role of soil and vegetation cover. We may, in fact, consider that this 'sea-cliff' system is really representative of the whole set of 'bedrock cliffs' and that, in other landscapes, such as system might occur *within* one of the major drainage

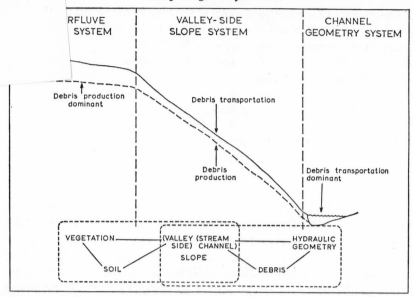

Fig. 2.22   The relationship between the systems of interfluves, valley-side slopes and stream channels

basins: for example as scree slopes within a glaciated valley. However, in the particular landscape with which we are concerned, the cliff-slope system is not only geographically but genetically separate from the drainage net.

As with the other two slope systems, it is only possible to obtain estimates of the 'characteristic' values of the major morphological variables by sampling. Since a cliff-line is a well-defined feature of limited extent, it is possible to sample either by selecting random 'eastings' (to determine the base of each profile) or by randomly initiating the set of samples and allowing the length of the first profile to determine the site of the second and so forth. We will assume that, since all the cliffs in the area are of the same lithology, formed by the same agency and facing in approximately the same direction, there are unlikely to be any significant thresholds within this subsystem. However, even so the form of the cliff itself may vary from a simple, sheer wall or 'free-face' to an 'arid' slope with a free-face above a 'constant' slope and the latter may be either bedrock or scree. The magnitude of the morphological system is consequently likely to vary from one sampling point to another but, by definition, the free-face subsystem should be present in all cases.

The major variables we might wish to include in such an analysis are:

1 The Strahler maximum angle (θ): i.e. the steepest non-cliff section.
2 The absolute maximum angle (β): i.e. the steepest cliff section.

3 The height of the profile (Ht).

4 The length of the profile (Lg).

5 The average angle of slope ($Ht/Lg = $ sine α).

6 The length of the Strahler maximum angle section (L_{SM}).

7 The length of the absolute maximum angle section (L_{AM}).

8 The proportion of the total slope length occupied by the Strahler maximum angle section ($L_{SM} \times 100/Lg = SM_L$).

9 The proportion of the total slope length occupied by the absolute maximum angle section ($L_{AM} \times 100/Lg = AM_L$).

10 The height/length integral $\left(\int_0^{100} dLg/dHt = H.L. \right)$.

11 The gradient of the shore at the base of the profile (γ').

12 The angle of apparent dip of the beds, as measured on the cliff-face ($+$ or $-\ \tau$): $+$ if the dip is outward, $-$ if it is inward.

13 The average spacing of vertical joints, as measured on the cliff-face (A_{jv}).

14 The average spacing of horizontal joints, and/or bedding planes, as measured on the cliff-face (A_{jh}).

15 The average lengths of the longest (Xa), intermediate (Xb), and shortest (Xc) axes of debris fragments.

16 The average number of principal faces of debris fragments (N_{pf}): this gives a direct measure of fragment shape.

17 The average degree of weathering of the fragments (W): this can only be estimated on an ordinal scale and will be more or less useful according to the lithology.

18a & 18b The average percentage vegetation cover of (a) the Strahler maximum angle section; and (b) the whole cliff (S_c; S_c').

Once again, there is considerable variation in the degree of accuracy to which these variables may be estimated: 1–10 are unique measurements, representing particular features of the cliff geometry, whereas 11–18b must all be derived from some form of sampling within the line of the profile. In addition, it has proved necessary to depart from the ratio scale of measurement for variable 17, and there are, in fact, several other characteristics of scree—such as its stability and degree of compaction—which would also have to be estimated in a similarly imprecise manner.

C. HYDRAULIC GEOMETRY

Several of the major features of the hydraulic geometry morphological system should be readily apparent: obviously thresholds are likely to exist between stream sections of different orders; there would, similarly, probably be differences between channels cut into different lithologies, but, as this area of study is of uniform lithology, the only possible source of such systematic noise would come from the difference between bedrock and alluvial reaches; finally, it is perhaps going to be difficult to decide upon a clear distinction between 'slope' and 'channel'.

The problem of sampling the stream reaches is best tackled by using the design for the valley-side slopes (Fig. 2.20), noting for each stream cross-section and the order of the channel, whether it is in alluvium or bed-rock and numbering it. A sample of each of the 10 possible subsystems of channel types may then be drawn at random. One point should be emphasised here: both valley-side slopes and channel systems may be categorised and sampled in the same operation and it would seem sensible to use *one* random sampling to define both subsystems, as it is then possible to study not only the morphology of each structure independently, but also the manner in which they are interlinked. This last step is one which has been frequently neglected by geomorphologists, for the very good reason that each system is itself highly complicated and their interactions, consequently, even more complex. However, this field of study is one which will have to be investigated further before we can really come to understand the way in which even the simplest of landscapes functions.

This close connection between slope and stream channel is frequently all too apparent when we come to try and define the latter in the field. Fig 2.23 shows some measurements of stream cross-sections and the lack of any definite break-of-slope. In some cases, however, clear banks exist and an upper limit to the normal confines of the stream can be set as the *bankfull stage*. By measuring the occurrence of discharges of differing magnitudes at such sections hydrologists and geomorphologists have concluded that a flood sufficient to fill the channel exactly to bankfull will be experienced once every 1·58 years (see Chapter 5 for a more detailed discussion). They have therefore tended to assume that, in cases where the physical limits of bankfull flows are difficult to determine, the "banks" should be set at the point to which the 1·58 years flood rises. This argument, it should be noted, is dangerous, if we are assuming that the physical event of bankfull flow is important: however, it does provide some arbitrary way of defining the stream channel.

Having located our cross-sections and delimited the bankfull limits, which variables should we consider as relevant to the hydraulic geometry morphological system?

1 The channel gradient (γ).
2 The upper channel width at bankfull (w).
3 The length of the wetted perimeter at bankfull (p).
4 The area of the cross section (a).

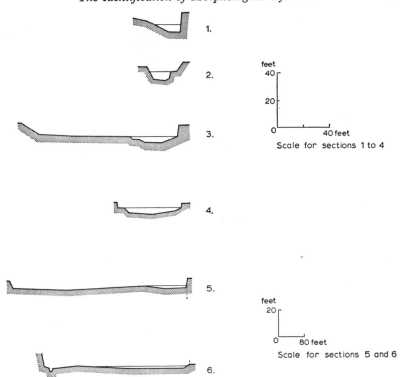

Fig. 2.23   Examples of stream-channel cross-sections in a semi-arid environment (southern Wyoming) illustrating the problem of deciding what constitutes the 'bankfull' stage

5 The hydraulic radius ($a/p = H/R$): this approximates 'mean channel depth'.

6 The bed roughness factor (Manning's n).

7 The average lengths of the three major axes of the bed material (Xa; Xb; Xc).

8 The average number of principal faces of the bed material (N_{pf}).

9 The angle of apparent dip of the bedrock ($+$ or $-$ τ).

10 The D_{84}, D_{50} and D_{10} values of the bed material.

11 The proportion, by weight, of silt and clay sizes particles in the bed material (SC_B).

12 The proportion by weight, of silt and clay size particles in the bank material (SC_S).

13 The proportion of the length of the wetted perimeter at bankfull occupied by vegetation (S_C).

6

It is rather unlikely that all these variables would be relevant at any one cross-section: for example, 7–9 would probably be of importance in torrential headwater reaches, while 10–12 would be of greater significance in more mature stretches. Here, then, is a case where the passage of a threshold in the system—that is, an increase in order and in discharge—causes a very marked alteration in the nature of the component variables. Of course, there is, in reality, no *abrupt* change from bedrock to alluvial channels and there will most likely be a group of cross-sections in which variables 7–12 will *all* be of importance.

D. BEACH MORPHOMETRY

In the section of the landscape with which we are concerned (Fig. 2.16) the mean high water mark happens to coincide with the base of the cliffs. The 'beach' therefore is easily defined as the area between the base of the cliffs and mean low water mark. This is, however, an unusually simple situation. Generally there is great uncertainty about the precise limits of the 'beach' system, because an area of 'storm beach' intervenes between the mean high-water mark and the coastline itself: it may well be that the storm beach is refashioned by each year's Spring tides, but in some cases it may be affected by only the most extreme storm surges. Where both a 'beach' and a 'storm beach' exist, they must be recognised as distinct elements of the landscape and their morphological systems investigated separately.

The extent of the beach, then, is from the mean high to the mean low water mark: a beach, however, is not merely a seaward slope, it is also a surface *across* which material is transported. It is, therefore, possible to view the morphological system of the beach in two ways: as a series of variables relating to profiles at right angles to the coastline; or as a set of attributes relating to *points* upon a continuous surface. Somewhat different variables are likely to prove important in each case.

Profiles

1 The maximum angle of slope (θ).
2 The length of the profile (Lg).
3 The height of the profile (Ht).
4 The average angle of the slope ($Ht/Lg = $ sine α).
5 The length of the maximum angle section (L_{SM}).
6 The proportion of the total profile length occupied by the maximum angle section ($L_{SM} \times 100/Lg = SM_L$).
7 The height/length integral $\left(\int_0^{100} dLg/dHt = H.L. \right)$.

8 The average D_{84}, D_{50} and D_{10} grain sizes of the material on the maximum angle section.

9 The average porosity or void ratio of the material on the maximum angle section (V_R).

10 The average penetrability of the material on the maximum angle section (P).

Points

1 The angle of slope seaward (θ).
2 The penetrability (P).
3 The D_{84}, D_{50} and D_{10} grain sizes of the material.
4 The porosity or void ratio (V_R).
5 The range of sand sizes (R_{SS}).
6 The moisture content of the sand (S_M).

Obviously the second approach will give a much more detailed picture of the *composition* of the beach, while the former is concerned to link the foreshore to the other major classes of slopes in the landscape.

E. CONCLUSIONS

Although the simplest of structures, morphological systems in the landscape are surprisingly difficult to delimit and it is equally tasking to identify all the variables which may be relevant in each case, bearing in mind the fact that, if one system contains a marked threshold, different factors may fill the same morphological niche on either side of the discontinuity. In the preceding discussion we have deliberately dealt with an extremely straightforward situation, in which thresholds are readily perceived. If we were to consider a *natural* landscape, however, with its almost inevitable variety of structure, lithology, inherited and contemporary landforms, the problem of identifying distinct subsystems would become infinitely more difficult.

This, then, is the first of the major difficulties: identifying the boundaries of systems and subsystems as exactly as possible, in order to eliminate systematic noise.

Our second difficulty is the variety of variables which may be involved: we have seen, for example that from some points of view we should include all the interfluve and all the hydraulic geometry variables with the valley-side slope subsystem. This proliferation of variables creates real problems. First, the more variables we include the more likely we are to make erroneous decisions concerning the nature of 'cause and effect' relationships. Second, the longer it is going to take to carry out our sampling programme. This second point sounds trivial, but it is really very serious and it is bound up with the difficulty of obtaining accurate estimates from samples. If we

A.

	D	θ	R	F	F₁	D₁	H
Drainage density (D)	1·00						
Maximum angle of valley side slope (θ)	−0·43	1·00					
Relative relief (R)		+0·625	1·00				
Channel frequency (F)	+0·94	−0·34	−0·15	1·00			
Frequency of first-order channels (F₁)					+0·97	1·00	
Desity of first-order channels (D₁)	+0·88					1·00	
Ruggedness number (H)		+0·78					1·00

B.

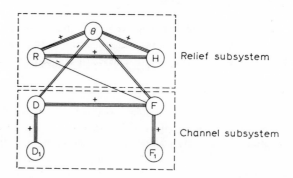

Relief subsystem

Channel subsystem

From Melton, 1958

Fig. 2.24

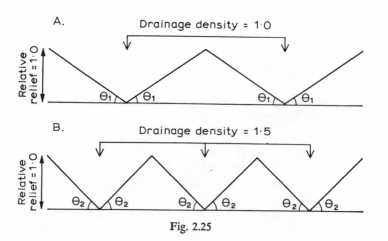

A. Drainage density = 1·0

Relative relief = 1·0

θ₁ θ₁   θ₁ θ₁

B. Drainage density = 1·5

Relative relief = 1·0

θ₂ θ₂   θ₂ θ₂   θ₂ θ₂

Fig. 2.25

wish to measure the angle of a section of a slope in the field there are several simple devices which will allow us to make a rapid estimate which should be within $\frac{1}{2}°$ of the true angle: it should not take us more than 10 or 15 minutes to arrive at a reasonably accurate estimate of the real angle, having made perhaps three different measurements. If we now wish to evaluate the average moisture content of the soil on that particular section of slope, we have *either* to take a great deal more than three measurements *or* to be content with a comparatively less accurate estimate. This is because soil moisture is likely to be constant neither throughout the upper layers of the soil nor along the line of the profile, largely because the soil itself will not be of exactly similar depth and type. It is equally true that the angle of slope is not precisely constant over the section, but our measurements are at a sufficiently large scale to allow us to disregard minor irregularities. When we come to measure soil moisture, we must take small samples and consequently, the likelihood that purely local variations will influence our final estimate becomes much greater.

This is the second major difficulty: that of the multitude of variables and the fact that it is considerably easier to obtain a good estimate of the true value of some than of others.

2.4 Some Descriptions of Morphological Systems

A. DRAINAGE BASIN MORPHOMETRY

This is in many ways the simplest of the morphological systems, if only because most of the variables may be estimated to a similar degree of accuracy. There have certainly been more studies of this system—or, to be accurate, of *parts* of this system—than of any other, except hydraulic geometry.

In general, people have looked not only at the general pattern of correlations between the variables involved but have tended to consider that *drainage density* $(D = \Sigma l_u/a)$ is the most dependent of the major factors. That is to say, the value of D is likely to be the product of the interaction of almost all the other features of the drainage basin. Although this is only partly true—because there are feedback loops in the system—it is, nevertheless a useful maxim.

Fig. 2.24*a* shows the matrix of significant correlations between a number of the major morphological features of over 80 4th order drainage basins in

Fig. 2.24 A simple correlation system for the main features of drainage basin form:
(*a*) The matrix of significant pairwise correlations
(*b*) The linkages within and between the relief and channel subsystems. (After Melton, 1958)

Fig. 2.25 An illustration of the relationship between drainage density and valley-side slope angle (θ), when relative relief is held constant

the western USA and Fig. 2.24*b* indicates the nature of the correlation structure. What is extremely clear is how far the purely geometric aspects of the relief and the stream network are interdependent: this is, of course, entirely to be expected, but the precise nature of some of the linkages is worth rather closer investigation. First, the maximum angle of valley-side slope is seen to be *inversely* related both to drainage density and channel frequency. This is in some ways predictable (in the Davisian view, as drainage efficiency increases, slopes decline) but in others surprising, for a denser network of rills should, on geometric grounds, produce a more accented topography, as Fig. 2.25 shows. At any event, this is a very clear feature of the system. It also produces a closed, *positive* feedback loop: $F + D - \theta - F$: i.e., the angle of slope declines irrevocably as the frequency of channels and the density of the drainage net increase.

Rather surprisingly, θ would seem to be a more 'critical' or 'nodal' variable than D, although none of its morphological relationships are at all unusual. The maximum angle of slope relates closely and directly to both the relative relief of the basin and the ruggedness (a dimensionless measure); inversely, and less directly, to overall drainage density and stream frequency and appears, as we have said, to decline irrevocably with the increasing development of the stream network. One final point to emerge is the importance of the joint variation in stream frequency and length, shown by the close positive association between D and F: this would seem to indicate that D increases by a proliferation of new channels, rather than by greater sinuosity of existing thalwegs (see Chapter 7.4).

Overall, this rather simple correlation structure would seem to provide an interesting starting point for the study of drainage basins. We must remember that it contains only 'static' morphological features so that we are forced, for genuine explanations, to look towards the cascading systems which may influence stream valleys, but the closeness of the links between these geometric variables should make us fairly confident that satisfactory explanations can be found.

B. SLOPE MORPHOMETRY

In view of the complexity of slope systems it is not surprising that no study exists which contains all those variables outlined in the second section of this chapter. As we shall see, the most complete analyses have been carried out of those structures where geometric variables are relatively more important than the non-geometric features of soil and vegetation cover, largely because it is a great deal easier to obtain estimates of the former characteristics.

Interfluve Slopes

Fig. 2.26 shows the significant linkages between five morphological variables characterising an interfluve slope system on sandy limestone in eastern France. It is very clear that the picture of the system which emerges is largely incomplete. The angle of slope appears to be inversely related to the degree of vegetation cover, which is weakly but positively associated with the weight of roots in the surface soil. Apparently quite separate as a subsystem, the depth of soil and the proportion of fine-grained material show a weak positive association (D_{10} is measured in grammes: the higher the value, the less fine-grained material present.)

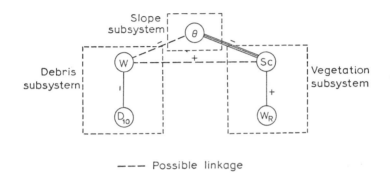

——— Possible linkage

After Kennedy, 1965

Fig. 2.26   The correlation structure of a simple interfluve slope system on sandy limestones in eastern France. (After Kennedy, 1965)

Although these correlations are of some interest, they do not really indicate the manner in which soil, vegetation and slope form are interrelated: two *possible* linkages are indicated in Fig. 2.26, but purely intuitively. Is there any reason why it should be more difficult to derive a satisfactory picture of the morphological system of interfluve slopes than that of drainage basin morphometry? The major difficulty is probably the very low variability of θ: most interfluve slopes are, by definition, rather gentle, this implies that those in any one area are likely to be rather similar in other respects, too, such as depth of soil cover, amount of vegetation and so forth. We are then, effectively using a coarse scale to measure slope and much finer scales to measure other variables: not surprisingly, this is not a very successful approach.

It would seem that it might be easier to define the pattern of links between soil, vegetation and slope in a situation where the last varies more widely.

A. Manitoba, streams towards (n=25)

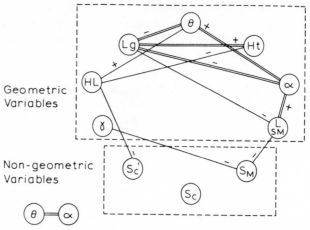

B. Manitoba, streams away (n=25)

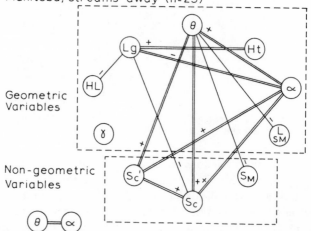

Fig. 2.27 The correlation structures of two discrete valley-side slope systems on sandy till in southern Manitoba, separated by a threshold in the intensity of basal fluvial erosion:

(a) The structure of a set of 25 slopes each of which has a stream close to its base. Note the lack of interlinkage between the variables representing slope and channel geometry and those describing the nature of the soil and vegetation cover

(b) The structure of a set of 25 slopes each of which has a stream moving away from its base. The interlocking of the non-geometric variables of soil and vegetation cover with the purely geometric features of slope and channel form is much better developed. (After Kennedy, 1969)

Valley-side slopes

Fig. 2.27 is based upon two sets of measurements of slopes cut in sandy till, in Manitoba: in both cases there is a much greater range of slope angles than in the study of the limestone interfluve profiles and the network of linkages, as predicted, is much more complete.

This pair of correlation structures is interesting in several ways:

1 If we take each structure as an example of a valley-side slope morphological system, then the most striking feature is the high level of association between the variables and, as predicted, the importance of the geometric features. Having said this, there are very few relationships which appear to be common to both systems, indicating that the position is by no means simple. There are, in fact, three common linkages: the maximum (θ) and mean (α) angles of slope are positively correlated; the height (Ht) and length (Lg) of profiles are positively correlated; and the length and average angle are negatively correlated. These three links express the fundamental geometry of slope forms and analysis of many other slope systems has shown that they are almost always present and strongly developed.

2 The low level of similarity between the two systems shown in Fig. 2.27 is in itself of interest, for these sets of profiles represent valley slopes which were thought to belong to different subsystems in that they lie on either side of a threshold in the local erosional environment. The first subsystem represents locations where the slope is being actively undercut by a stream: the second consists of the opposing valley sides or the corresponding slip-off profiles. The difference in the nature of the two morphological systems would seem to indicate that this threshold in erosional activity is of real importance to slope form. If we analyse the variation between the two systems, the clearest difference lies in the relative number of linkages between the geometric and non-geometric variables. In the case of the undercut slopes most of the significant correlations are within the geometric sub-set, implying that the most important feature of these profiles is the rather rapid transportation of debris to the stream. In the case of the slip-off slopes, on the other hand, there are numerous strong bonds between the soil and vegetation characteristics and the angular features of the profiles, in particular: this would seem to suggest that here the tendency is for a more stable debris-*production* (i.e. weathering) situation.

3 Within the broad distinction in morphological systems between the undercut and slip-off profiles there are one or two specific correlations which are of general interest. For example, the irreversible reduction of relief through time appears in Fig. 2.27a as the set $\theta \longrightarrow \alpha \longrightarrow Lg \longrightarrow \theta$ i.e. slopes lengthen

and flatten continuously. There is also a positive feedback situation in Fig. 2.27 between $\theta + \alpha + S_c' + \theta$ suggesting that the degree of vegetation cover *increases* as the average angle of the profile increases. This latter situation is, quite patently, impossible: when θ reaches a critical value, S_c' will begin to decline until, with vertical cliffs, one would expect to find no vegetation cover at all (this negative correlation appears in the interfluve system shown in Fig. 2.26). It appears that the direction of association between slope angle and vegetation cover is not stable, as are the geometric relationships, but that it varies with the nature of the particular slope subsystem.

This pair of correlation structures indicates not only something about the way in which the till slopes of southern Manitoba are organized but, more important perhaps, points up several general conditions concerning valley-side slopes the major one being the 'instability' of the form and direction of many such morphological linkages.

Cliff systems

Fig. 2.28 shows the correlation diagrams of two sets of cliffs from the English Lake District, which have been separated because they are of different lithology. It is clear that, although there are some basic similarities between the slate and volcanic cliffs (Figs. 2.28*a* and 2.28*b*) there are also very substantial differences.

The major similarity is in the close association of the three measures of debris size (Xa, Xb, Xc), further, the positive correlation between Xa and the angle of the cliff-face (β), which suggests a tendency for larger blocks to fall from steeper cliffs. There is, however, a difference in the relation between Xc and β: on the slate cliffs this is negative, suggesting that blocks falling from steep cliffs are rather slab-like, while the positive correlation for the volcanic slopes implies that material breaks into more rectangular fragments.

More interesting than these similarities are the differences between the two systems. The slate cliffs show a much more clearly defined set of inter-relationships than the volcanics and, on the whole, would seem to suggest a system which is in a rather nicer stage of internal adjustment than that of the profiles on volcanic rocks. This is rather difficult to prove from a morphological study, but the fact that the angle and length of the cliff section of the former slopes are related to the length of the scree slope, as well as to the size of the scree debris would seem to be a useful indicator, as would the rather close association between debris size and shape, degree of weathering and amount of vegetation cover of the scree slopes. If we were to summarise the position, we could perhaps say that the slate cliffs give more evidence of

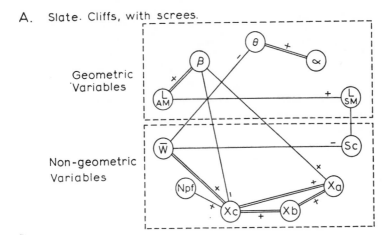

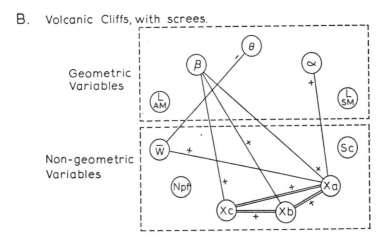

Fig. 2.28   The correlation structure of two discrete cliff systems in the Lake District of the British Isles, separated by a threshold in lithology (i.e. resistance to erosion)

(*a*) Illustrates the relative separation of geometric and non-geometric subsets on the unstable, slate cliffs

(*b*) Shows the slightly closer linkages between the two sets, but the lesser overall degree of 'organization' on the 'moribund' volcanic cliffs. (After Towler, 1969)

being in *dynamic equilibrium* at the present time than do the volcanic slopes (see Chapter 6).

Summary

All three studies of slope systems illustrated have included both geometric and non-geometric variables and it seems that the linkages within the former categories emerge most clearly in the case of valley-side profiles, whereas those among the latter are most readily defined for scree slopes. In all cases, morphological links between slope geometry and mantle *are* discernible but it is apparent that the form and, more important, the *nature* of these correlations is highly dependent upon the precise type of subsystem studied.

C. HYDRAULIC GEOMETRY

In most cases hydraulic geometry has been studied as a process-response system, with the morphology of the stream channel being viewed in the light of changes in discharge. One can, however, gain something from considering the features of stream cross-sections themselves and this is particularly true in the case of the ephemeral stream channels of semi-arid areas. Fig. 2.29 illustrates the morphological systems of samples of 4th (*A*) and 2nd (*B*) order ephemeral channels in an area of sandy conglomerate in Wyoming.

It is clear that variation in stream order does have a very marked effect upon the nature of the morphological system and equally clear that the smaller streams show the tighter pattern of interrelationships. The reason for this would seem to lie in the greater importance of the size of bed material in the case of the smaller channels, as this is closely connected with all the other features of channel geometry (Fig. 2.29*b*). It is probable that in small channels the material has travelled only a short distance from its source, so that the size of debris is effectively a 'given' feature of the system and channel shape must be adjusted accordingly. In larger or more mature systems, however, debris may have travelled further and been both sorted and reduced in size so that its *direct* importance on channel form is lessened.

In this case, we obviously cannot interpret the difference in the closeness of linkages between 2nd and 4th order systems in terms of a greater or lesser approach to dynamic equilibrium. The less well 'organized' situation in the latter case is produced by the decline in importance of the particular debris characteristics selected for the study and a failure to replace them by other measures of bed roughness—for example, Manning's 'n'—which might prove more appropriate to the subsystem of the larger channels.

A. (4th order)

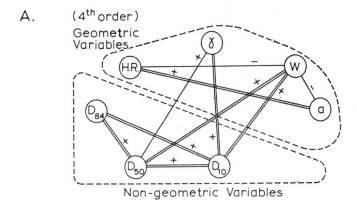

B. (2nd order)

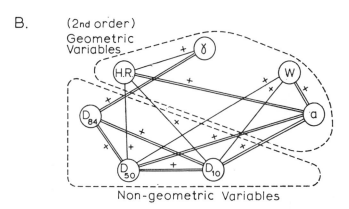

Fig. 2.29   The correlation structures of two discrete stream channel systems on sandy conglomerates in southern Wyoming, separated by a threshold in stream magnitude:

(*a*) The structure for a set of 4th order channels, showing only a moderate degree of interlock between the features of channel shape and particle size

(*b*) The much more highly interlinked structure of a set of 2nd order channels

D. BEACH MORPHOMETRY

Fig. 2.30 illustrates a rather different form of study of a morphological system, in this case an analysis of a group of at-a-point beach properties. Penetrability (*P*) has been considered as the dependent variable in the system and individual logarithmic correlations have been calculated between *P* and mean sand diameter, range of sand sizes, moisture content of the sand and sand porosity. These correlations show that grain diameter is the single most important variable, explaining 69·79% of the observed variation in *P*. However, the high degree of interlock between all the variables (shown very clearly in Fig. 2.30*b*) is not only supported by theoretical considerations but also by the fact that the assumption of addivity gives a total explanation of 125·15% (i.e. 33·92% + 20·48% + 69·79% + 0·96%). Clearly there is considerable replication in the system and its nature begins to emerge when values of R² are calculated for all pairs of independent variables (except those containing porosity which is obviously quite unimportant).

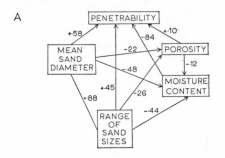

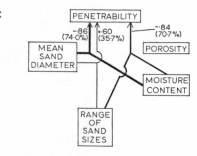

Fig. 2.30   Relationships between the factors controlling beach firmness. (Data from Krumbein, 1959):

(*a*) The complete correlation (*r*) linkages between all the variables, together with coefficients of correlation and suggested directions of control.
(*b*) The correlation (*r*) linkages significant at the 95% (single arrow), 99% (double arrow) and 99·9% (triple arrow) levels.
(*c*) Coefficients of multiple correlation and percentage explanations (i.e. R^2) between three pairs of independent variables and beach penetrability.

    Combining the effect of size range with mean diameter gives an explanation only 1·80% greater than that achieved by the latter variable alone (i.e. 35·72% over 33·92%), rather than the additive 54·40% (i.e. 33·92% + 20·48%). There must, then, be considerable repetition of information between mean diameter and size range and it is clear that the small grain sizes and small size ranges commonly occur together. In a similar fashion, grain size only adds 4·21% (i.e. giving 74·00% over 69·79%) when combined with moisture content—far short of its individual contribution, so that it is apparent that the amount of moisture held in sand is very much a function of its mean grain size. However, the latter variable is shown to be the second most important control of beach penetrability and Fig. 2.31 illustrates

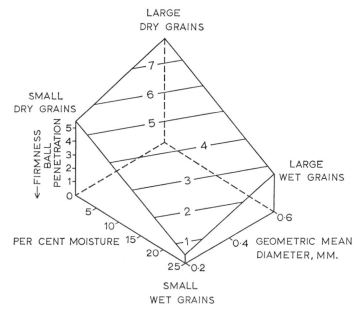

Fig. 2.31   Linear component trend surface of beach (ball) penetrability
(Y) as a function of (geometric) mean sand diameter (X_1) and moisture
content (X_3)

The equation of this surface is:

$$Y_C = 4{\cdot}603 + 5{\cdot}286X_1 - 0{\cdot}024X_3. \quad (R_{13}{}^2 = 0{\cdot}74)$$

(After Krumbein, 1959)

graphically the fact that large, dry grains give greater penetrability than
small, wet ones.

From here, the analysis proceeded to combine mean grain size, size range
and moisture and this step shows that all three together explain more than
76% of the variation in P, leaving the question of what other factors (not
built into this study) are important in accounting for the remaining 24% of
the observed variation in beach penetrability. Finally, Fig. 2.32 shows the
result of a linear surface fitted to represent the combined effect of medium
sand size and range of sand sizes on the slope of the foreshore (θ), indicating
that, other things being equal, coarse well-sorted sand can be expected to
give the steepest foreshore slope. Poorly sorted sand (i.e. consisting of a
wide range of sizes) might theoretically be expected to support steeper slopes
than well-sorted sand, but this effect is probably more than counteracted
by the higher permeability of the latter which reduces the amount of
backwash and inhibits the amount of material washed down the beach.

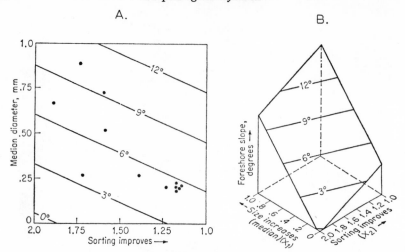

Fig. 2.32   Linear component trend surface of beach foreshore slope (Y) on Clark Street Beach, Evanston, as a function of median grain diameter (X_1) and degree of sand sorting (X_2) (From Krumbein and Graybill, 1965):

    (*a*) Graph of observed data together with the linear trend surface:

$$Y_c = 8{\cdot}31 + 11{\cdot}48X_1 - 4{\cdot}62X_2 \ (R_{12}{}^2 = 0{\cdot}61)$$

    (*b*) Block diagram derived from (*a*)

2.5 Conclusion

The identification, description and interpretation of morphological systems are fundamental to all physical geography, whether or not this is done in any highly sophisticated fashion. Our major difficulty is that there are so many different morphological systems in even the smallest section of a real landscape and that the thresholds between them are by no means always readily apparent. It must be stressed again that it is vital to discover not only the position of such thresholds but also the way in which they affect the nature of the system: Does the intensity of a correlation change, or does the direction of association reverse? Is the importance of one variable reduced or is its place taken by another? Until we can find satisfactory—and that means accurate—answers to these questions, studies of more complex, natural systems are bound to be at a rather superficial level.

REFERENCES

Blalock, H. M. (1964), *Causal Inferences in Nonexperimental Research* (Univ. of North Carolina), 200.

Blalock, H. M. (1969), *Theory Construction* (Prentice-Hall, New Jersey), 180.

Chorley, R. J. and Haggett, P. (1965), Trend-surface mapping in geographical research; *Transactions of the Institute of British Geographers*, **37**, 47–67.

Chorley, R. J. (1966), The application of statistical methods to geomorphology; In Dury, G. H. (Ed.) *Essays in Geomorphology* (London), 275–387.

Coates, D. R. (1958), Quantitative geomorphology of small drainage basins of southern Indiana; *Office of Naval Research, Project NR 389–042, Technical Report, 10.*

Croxton, F. E. and Cowden, D. J. (1948), *Applied General Statistics* (Prentice-Hall, New York), 944.

Ezekiel, M. and Fox, K. A. (1959), *Methods of Correlation and Regression Analysis* (Wiley, New York), 548.

Fisher, R. A. (1960), *The Design of Experiments* (Oliver and Boyd, Edinburgh), 7th Edn., 248.

Gregory, S. (1962), *Statistical Methods and the Geographer* (Longmans, London), 240.

Kennedy, B. A. (1965), *An Analysis of the Factors Influencing Slope Development on the Charmouthien Limestone of the Plateau de Bassigny, Haute-Marne, France*; (B.A. Dissertation, Department of Geography, Cambridge University), 99.

Kennedy, B. A. (1969), *Studies of Erosional Valley-Side Asymmetry*; (Unpublished Ph.D. Thesis, Cambridge University), 289.

King, L. J. (1969), *Statistical Analysis in Geography* (Prentice-Hall, New Jersey), 288.

Krumbein, W. C. (1959), The 'sorting out' of geological variables illustrated by regression analysis of factors controlling beach firmness; *Journal of Sedimentary Petrology*, **29**, 575–587.

Krumbein, W. C. and Graybill, F. A. (1965), *An Introduction to Statistical Models in Geology* (McGraw-Hill, New York), 475.

Maxwell, J. C. (1960), Quantitative geomorphology of the San Dimas Experimental forest, California; *Office of Naval Research, Geography Branch, Project NR 389–042, Technical Report 19.*

Melton, M. A. (1957), An analysis of the relations among elements of climate, surface properties and geomorphology; *Office of Naval Research Project NR 389–042, Technical Report* 11 (Department of Geology, Columbia University, New York), 102.

Melton, M. A. (1958A), Geometric properties of mature drainage systems and their representation in an E_4 phase space; *Journal of Geology*, **66**, 35–54.

Melton, M. A. (1958B), Correlation structure of morphometric properties of drainage systems and their controlling agents; *Journal of Geology*, **66**, 442–460.

Melton, M. A. (1960), Intravalley variation in slope angles related to microclimate and erosional environment; *Bulletin of the Geological Society of America*, **71**, 133–144.

Morisawa, M. E. (1959), Relation of quantitative geomorphology to stream flow in representative watersheds of the Appalachian plateau province; *Office of Naval Research, Geography Branch, Project NR 389–042, Technical Report 20.*

Olson, E. C. and Miller, R. L. (1958), *Morphological Integration* (University of Chicago Press), 317.

Strahler, A. N. (1964), Quantitative geomorphology of drainage basins and channel networks; In Chow, V. T. (Ed.), *Handbook of Applied Hydrology* (McGraw-Hill, New York), Section 4–11.

Towler, J. E. (1969), *A Comparative Analysis of Scree Systems Developed on the Skiddaw Slates and Borrowdale Volcanic Series of the English Lake District*; (B.A. Dissertation, Department of Geography, Cambridge University), 84.

3: *Cascading Systems*

> You cannot put your hand in the same stream
> twice—No, not even once!
>
> CHINESE PHILOSOPHER (CHO LEE?)

3.1 The Nature of Cascading Systems

Cascades are one of the most important types of dynamic system and are defined as structures within which the *output* from one *subsystem* forms the *input* for the next subsystem and within which a *regulator* may operate either to divert a part of the input of mass or energy into a *store* or to create a *throughput*, producing the subsystem output.

These basic components of cascading systems may be represented as *canonical structures*. The most important symbols for illustrating these are given in Fig. 3.1 and they are combined into a simple canonical structure in Fig. 3.2. Here an input $(A + B + C)$ into Subsystem I is operated on by two regulators in succession: the first placing part (A) into store from which it is subsequently output; the second producing two separate outputs (B and C). The combined output $C + A$ cascades as input into Subsystem II, while output B forms a feedback to the source of the original input $(A + B + C)$. From this example it should be clear, firstly, that the regulators occupy key positions in controlling the operation of the cascade and, secondly, that the processing associated with storage may introduce an important time lag in the flow.

Several fundamental points concerning the nature of cascades should, perhaps, be made here:

(*a*) Most cascades of interest to the physical geographer ultimately obtain their energy from the major terrestrial cascading system: the solar energy cascade.

(*b*) As the receipt of radiation from the sun varies continuously, so, too, does the magnitude of energy inputs into all the connected cascades. This variation in energy input is of importance since it may be of considerable magnitude in some cascades, as we shall see. It also places a premium upon the accurate measurement of energy flows within the major cascading systems, a condition which is often very difficult to meet, with the result that many analyses must rely upon rather basic observations of inputs and outputs, the nature of the internal transfers and storages being simulated by computer programming.

77

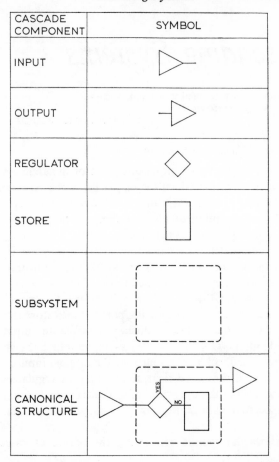

CASCADE COMPONENT	SYMBOL
INPUT	
OUTPUT	
REGULATOR	
STORE	
SUBSYSTEM	
CANONICAL STRUCTURE	

Fig. 3.1   The symbols used in depicting the canonical structures of cascading systems

(c) All the cascades of interest to the physical geographer interlock, at some point, with variables in morphological systems such as those described in Chapter 2. The interaction of cascade and morphological systems produces the process-response system, which is discussed in Chapter 4.

(d) The role of the regulators within cascades—frequently fulfilled by morphological variables—is of particular relevance to the geographer, as this very often represents a key point at which Man can intervene to change or control the natural system. This topic will be explored further in the final chapter.

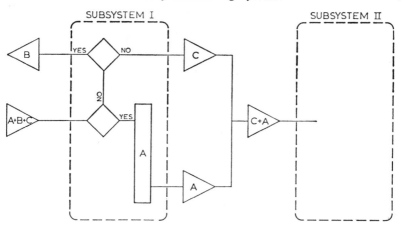

Fig. 3.2   A simple canonical structure of a cascading system involving two subsystems

3.2 The Major Cascading Systems

A. INTRODUCTION

Table 3.1 gives a list of the main subsystems of the solar energy cascade which are employed in this work. It is clear at once that, despite their varied magnitude and composition, each of them represents a specific geographical environment (i.e. one having magnitude and location), linked to adjacent environments by outputs and inputs. Thus the major subsystems range from the atmosphere and the ocean at one end of the magnitude spectrum, to the soil subsystem or that of grass cover and its interstitial air, near the other extreme. The two major cascades of interest to the geographer are that of solar energy and its important component, the hydrological cycle. It is clear that the lesser subsystems of the atmosphere and the vegetation cover are components of the former cascade and that the vegetation and soil subsystems form subsystems of the latter. In the same way water and sediment cascade from the deep water subsystem, though those of the shoaling and breaker zones, into the swash zone; and debris provided by the cliff subsystem cascades into that of the scree.

Table 3.2 lists the major regulators which are identified in this book: these can be divided into three groups. Firstly, there are the 'threshold regulators' (such as surface infiltration capacity) which control storage decisions concerning the energy, water or debris entering a subsystem. Secondly, there are 'dispositional regulators' which, although not presenting such obvious thresholds, nevertheless control the disposition of energy or mass: for example, will one particular input of water be subject to evapotranspiration, or not? Thirdly, there are 'presence or absence regulators',

TABLE 3.1   SPATIAL SUBSYSTEMS

Atmosphere

Lowest atmosphere

Earth

Land   Ocean

Slope   Cliff

Upper scree   Lower scree   Basal slope

Channel/Valley

Stream channel

Vegetation

Canopy

Trunk space

Crop   Cotton   Grass

Surface   Swash zone

Soil   Breaker zone

Aeration zone   Shoaling zone

Ground-water zone   Deep water zone

such as whether a stream is present at the base of a valley side or not. It is clear that many of these regulators play important feedback roles by linking cascades to morphological structures, thus forming process-response systems. It should also be emphasised that many of these regulators—particularly the threshold regulators—are extremely susceptible to human intervention: infiltration capacity is an all too familiar example. The regulators, then, provide an important group of *valves*, and by manipulating these the process-response system is transformed into a control system.

After passing the regulators, energy and mass may be diverted into stores, shown in Table 3.3. These stores are of three types:

1 Mass storage. This includes the storage of water (for example, as ground water) and that of debris: the latter may be either positive (for example, the accumulation of debris at the base of a slope) or negative (as in channel erosion).

TABLE 3.2 REGULATORS

THRESHOLDS

Exceeds angle of repose?	$>A^*$
Exceeds limiting interception storage?	$>I^*$
Exceeds infiltration capacity?	$>F^*$
Exceeds limiting surface storage?	$>R^*$
Exceeds soil field moisture capacity?	$>M^*$
Exceeds limiting areation zone storage?	$>L^*$
Exceeds channel storage capacity?	$>S^*$

DISPOSITION OF ENERGY OR MASS

Absorbed?	a?
Reflected or scattered?	r?
Evapotranspired?	e?
Intercepted?	i?
Penetrates?	p?
Trapped by vegetation?	v?
Friction loss?	f?
Thermal dissipation?	t?
Change in energy?	c?
Long wave radiation?	l?

PRESENCE OR ABSENCE

Basal stream present?	*BS*?
Coarse material available?	*CM*?
Within drainage basin?	*WB*?
Ocean surface present?	*OS*?
Land surface present?	*LS*?

TABLE 3.3 ENERGY AND MASS STORAGES

MASS

Water:
Interception storage	I
Surface detention storage	R
Soil moisture storage	S_M
Beach moisture storage	M'
Aeration zone storage	L
Ground-water storage	G
Flood-plain storage	F
Channel storage	S

ENERGY

Radiation absorbed by clouds (C_a)
Radiation absorbed by air (A_a)
Heat
Heat transport
Change in heat storage
Energy change (ΔE)
Change in soil heat (ΔG)

Debris:

Increase: Scree accumulation
Basal slope debris accumulation
Channel debris $\begin{cases} \text{Suspended load amount} \\ \text{Bed load amount} \end{cases}$

Decrease: Slope erosion
Channel erosion
Shoreward transport
Debris transport

SOME CONTROLS OVER MASS AND ENERGY GENERATION

Cliff height   (H_c)
Cliff angle   (β)
Susceptibility of cliff to weathering
Wave period (T)
Wave angle (α)
Wave height (H_0)

2 Energy storage. This is often much more difficult to establish, but it includes absorbed radiation (for example, by clouds), heat transport and changes in heat storage. (A principal area of energy storage is, of course, as chemical energy in plant and animal tissues, but the significance of this in terms of those systems primarily considered in this book is probably marginal.)

3 Some indirect controls over the quantities of mass and energy generated within subsystems. These involve such parameters as those describing wave geometry (and hence energy) and those concerning the geometry and susceptibility to weathering of cliffs supplying debris to adjacent scree subsystems.

TABLE 3.4 INPUTS AND OUTPUTS OF MASS AND ENERGY

Mass

Water: Precipitation amount/intensity—p
Relative January precipitation intensity—J
Precipitation effectiveness—$P–E$
Channel precipitation—P_J
Evapotranspiration—e
Evapotranspiration from vegetation—e_i
Evapotranspiration from surface—e_r
Evapotranspiration from soil—e_m
Evapotranspiration from aeration zone—e_e
Channel evaporation—e_c
Stemflow and drip—i
Unconcentrated surface runoff—q_0
Stream channel runoff amount/intensity—q
Infiltration into soil or beach—f
Throughflow—m
Downward percolation from soil—s
Interflow—l
Deep percolation to ground water—d
Ground-water flow—g
Deep basin outflow—b
Offshore seepage—b'

Energy

Solar: Total incoming solar radiation—$Q_s\ (=Q + q')$
Direct beam radiation—Q
Diffuse radiation—q' or q
Ground-level radiation
Evapotranspiration energy—LE
Sensible heat radiation—H
Long-wave radiation—I_L
Earth's surface albedo—$(Q + q')\alpha$
Energy reflected by vegetation—C_v
Energy reflected by clouds—C_r
Energy reflected by air particles—a_r
Heat conducted

Stream and Wave: Energy from precipitation and stream downcutting—E_i
Energy output at channel mouth—E_0
Energy dissipation due to radiation, evaporation, chemical energy, etc.—E_d
Wave energy
Modified wave energy (by friction losses)
Swash energy

Debris:

Debris supply $\begin{cases} \text{Debris fall} \\ \text{Rolling and sliding debris} \\ \text{Creeping debris} \end{cases}$
Water and debris runoff

The inputs and outputs of energy (Table 3.4) clearly involve a large number of different variables, which may be divided into two classes: those primarily

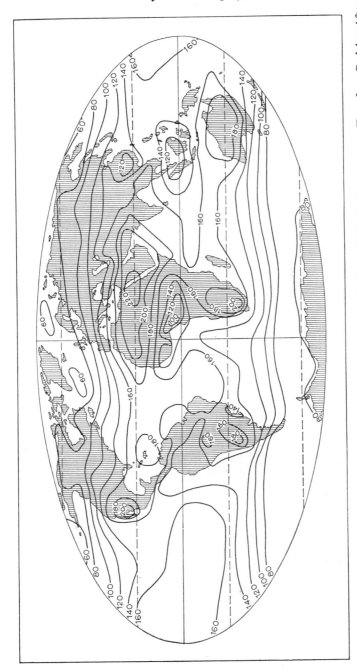

Fig. 3.3   The average annual solar radiation on a horizontal surface at ground level, in kilolangleys per year. (Data from Budyko. After Sellers, 1965)

Cascading Systems

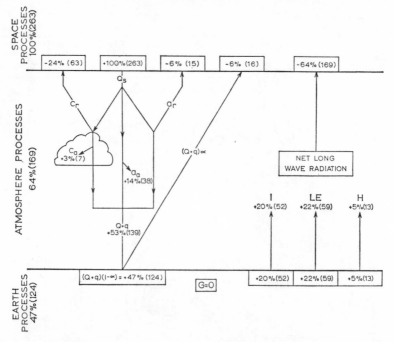

Fig. 3.4 The solar energy cascade in diagrammatic form. Values are given in per cent of annual totals and (in parenthesis) in kilolangleys per year. (Data from Budyko and others). For explanation of Symbols, see Table 3.4 and p. 88

related to mass, and those purely to do with energy. Inputs and outputs of mass in those systems of interest to the physical geographer may involve water (for example, stream channel runoff) or the rate of supply of debris. Inputs and outputs of energy involve, among others, solar energy (for example, radiation of sensible heat) or energy associated with stream or wave action (for example, deep water wave energy).

To summarise. The cascading systems with which we are dealing differ considerably in magnitude, but are all ultimately interconnected. The regulators controlling these cascades can be of several types, of which the most important are the threshold regulators since these determine the extent of storage of both mass and energy. Clearly, the study of cascading systems not only demands a knowledge of the amount of mass and energy transferred from one subsystem to another, but also involves the more difficult problem of the manner in which such cascades link with morphological structures. We must now turn to the principal cascades and examine their structure in some detail.

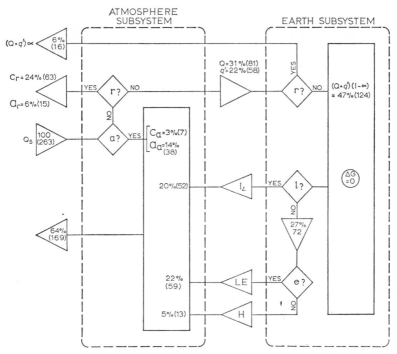

Fig. 3.5   The solar energy cascade shown in Fig. 3.4 depicted as a canonical structure

B. THE SOLAR ENERGY CASCADE

This is, as we have said, the most important and fundamental of the terrestrial energy cascades and most other energy cascades of relevance to the physical geographer are ultimately derived from it. For example, 92·5% of the observed differences in stream discharge between five neighbouring experimental basins in West Virginia could be explained by differences in the calculated potential insolation. That it is not inappropriate to view the input of solar energy as a cascade is illustrated by the simile used by a meteorologist working at high elevations, who wrote: 'The heat load in the alpine tundra advances like an avalanche as the sun bursts upon the mountain slope'. Solar energy is commonly expressed in units of the *langley* (ly), where 1 langley is 1 gram-calorie/cm² and a *kilolangley* (kly) is 1000 langleys.

Although the amount of solar energy received at the earth's surface does vary markedly from place to place (Fig. 3.3), it is convenient to examine the gross terrestrial energy budget as if it were two interlocking cascades: the one of incoming short-wave radiation; the other of outgoing long-wave radiation. Fig. 3.4 expresses these cascades in diagrammatic form and Fig. 3.5 as a canonical structure. Stated in simplified form, if the total

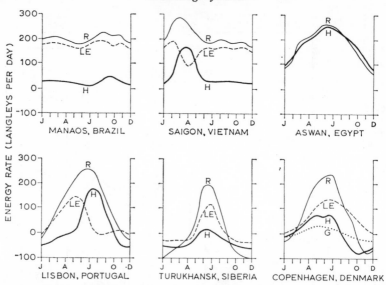

Fig. 3.6　　The annual variations of the energy balance components (langleys per day) at various geographical locations. In the first five of these G is assumed to be zero. (Data from Budyko. After Gates, 1962). For Copenhagen the values of G are indicated. (After Sellers, 1965). For a definition of R, LE, H and G, see p. 88

predominantly short-wave solar radiation at the top of the atmosphere (Q_s) is 263 kly/yr (100%), 63 (24%) are reflected and scattered back into space by clouds (C_r)† and 15 (6%) by air molecules, dust and water vapour (A_r). Clouds (C_a)† and air (A_a) absorb, and are heated by, 7 kly/yr (3%) and 38 (14%) of incoming radiation, respectively; whereas 139 (53%) reach the earth's surface by direct beam [$Q = 81$ kly/yr (31%)] and diffuse solar radiation [$q = 58$ kly/yr (22%)]. The earth reflects 16 kly/yr [$(Q + q)\alpha = 6\%$] back into space, and absorbs 124 [$(Q + q)(1 - \alpha) = 47\%$]. The mean reflectivity of the surface, the *albedo* (α), is 0·115 (i.e. on average, 11·5% of the solar radiation *reaching the earth's surface* is reflected); but this is naturally very variable being as high as 0·950 for fresh snow and as low as 0·020 for high-angle rays on clear water.‡ Of the incoming short-wave radiation cascade, the earth subsystem processes some 124 kly/yr (47%) and the short-wave radiation balances can be expressed as:

$$Q_s = C_r + A_r + [(Q + q)\alpha] + C_a + A_a + [(Q + q)(1 - \alpha)]$$

† These gross figures are naturally subject to very great geographical and seasonal variation.

‡ Other values of α are: Dark soil 0·050–0·150; Deserts 0·250–0·300; Deciduous forests 0·100–0·200.

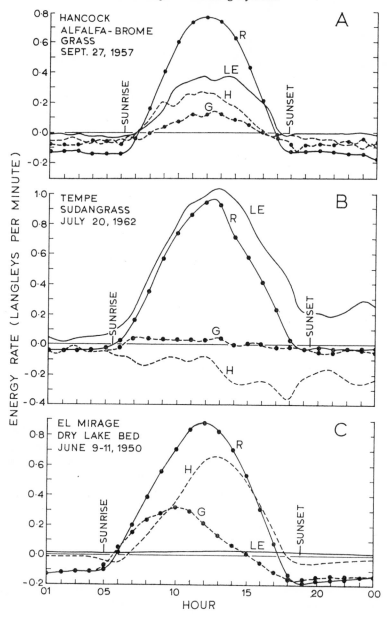

Fig. 3.7   The diurnal variations of the energy balance components (langleys per minute) over grass at (a) Hancock, Wisconsin, and (b) Tempe, Arizona; and (c) over the virtually bare surface of a dry lake bed at El Mirage, California. (From Sellers, 1965). An explanation of the symbols is given on p. 88

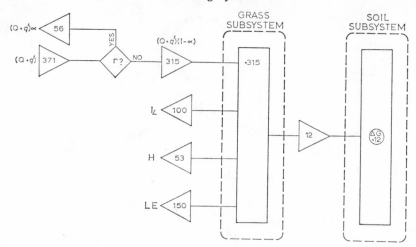

Fig. 3.8   Cascade of solar energy (langleys per day) on a meadow near Hamburg during May 1958. (Data from Miller, 1965)

The long-wave radiation balance between the earth's surface and atmosphere (*R*) is expressed by:

$$R = LE + H + G$$

and the total surface radiation heat loss is $R + I$: where LE is the transfer of latent heat by evapotranspiration (59 kly/yr = 22%), H is the transfer of sensible heat conducted from the earth and convected in the atmosphere (13 kly/yr = 5%), G is the change in surface (soil) heat store (assumed to be zero if the earth is in thermal equilibrium), and I is the long-wave radiation from earth to atmosphere. Clearly, the absolute values of these components vary greatly with season and geographical location (Fig. 3.6) and with time of day (Fig. 3.7). The long-wave balance is maintained by a net radiation from the atmosphere into space of 169 kly/yr (64%), and it is evident that this amount represents the energy processed by the atmosphere subsystem (7 + 38 + 52 + 59 + 13 = 169 kly/yr).

The above general view of the solar energy cascade must obviously be greatly modified when one considers the energy exchanges in different geographical locations, particularly where there are differences in the character of the surface. Fig. 3.8 shows the average daily exchanges involving the grass and soil subsystems for a Hamburg meadow during May 1958, and Fig. 3.9*a* and *b* a more detailed energy cascade involving a short green crop in southern England for the period May–September 1949. The latter example shows that only some 1% of the incoming energy is used in the actual growth of the crop and even for forests the heat required for growth

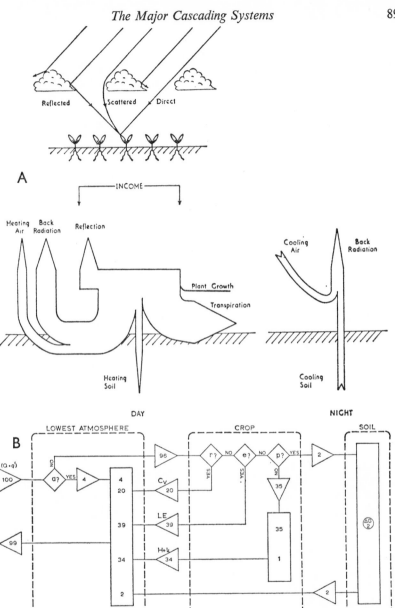

Fig. 3.9   Solar energy balance for a short green crop in southern England during May to September, 1949

(a) Schematic representation of the solar energy balance, showing daytime and night conditions. (From H. L. Penman, The Calculation of Irrigation Need; *Min. of Agr., Tech.* Bull. 4)

(b) Canonical depiction of the date from (a). The numbers are percentages

A.

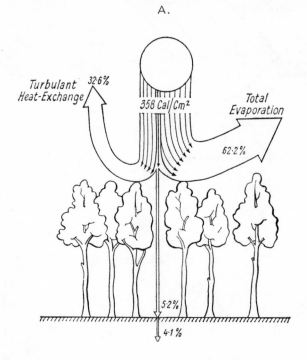

B.

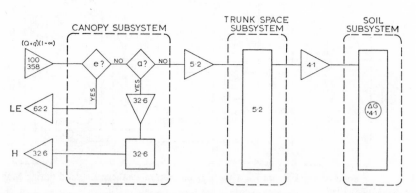

Fig. 3.10   Solar energy cascade on an average day during June to August in a 30-year-old oak stand in the Tellerman Experimental Forest, Voronezh Province, U.S.S.R.

(*a*) Schematic representation of the cascade. (From Sukachev and Dylis, 1968)
(*b*) Canonical depiction of the data (per cent) from (*a*)

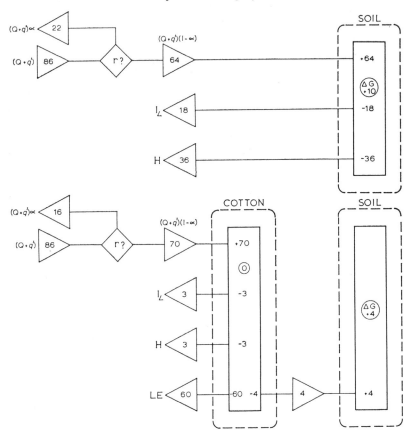

Fig. 3.11   The solar energy cascade at noon during July 1953 at Bukhara, Central Asia (langleys per hour). (Data from Miller, 1965)

Above. In the bare desert

Below. In a nearby irrigated cotton field

is relatively small—a dense stand of Scots Pine utilises only some 1 kly/yr (of which 40% produces forest floor litter and 60% wood tissue) and tropical forests only 1·2 kly/yr. A simplified view of the incoming solar energy cascade through the canopy, trunk-space and soil subsystem of a 30-year experimental oak forest (crown density 0·9–1·0) in Voronezh Province, U.S.S.R. on an average day during June–August is given in Fig. 3.10 *a* and *b*. An extreme example of the effect of a vegetation cover on the incoming solar energy cascade is provided by recordings over the open desert and over nearby irrigated cotton fields at Bukhara in Central Asia at noon on a July day in 1953 (Fig. 3.11).

A.

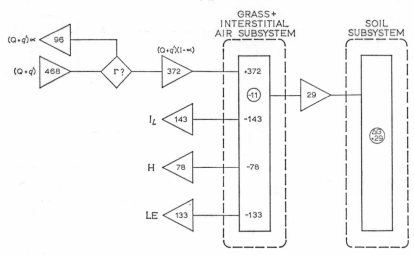

B.

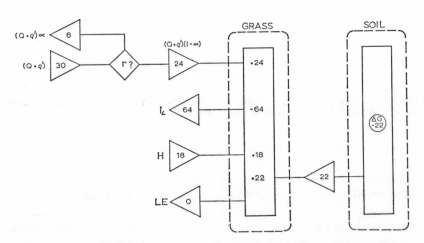

Fig. 3.12   Seasonal variations in the solar energy cascade in a grass plot near Copenhagen (langleys per day). (Data from Miller, 1965)

(*a*) During a June day, having 17 hours of daylight from a sun reaching an altitude of
58°

(*b*) During a December day, having 7 hours of daylight from a sun reaching an altitude
of 11°

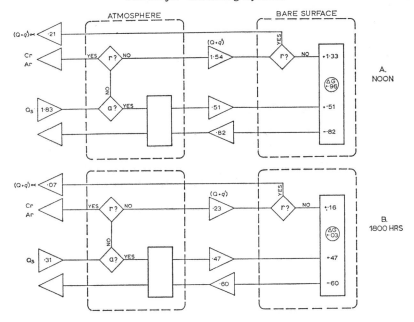

Fig. 3.13   Diurnal variations in the solar energy cascade over a bare rock surface at 5280 feet elevation in Blue Canyon, California, on 18 July, 1963 (langleys per minute). (Data from Miller, 1965)

(*a*) At noon with the sun at an altitude of 70°. The surface temperature of the rock is 52°C. (*b*) At 1800 hours with the sun at an altitude of 10°. The surface temperature of the rock is 22°C

Seasonal variations in the solar energy cascade are usually very striking, particularly where pronounced vegetational changes are involved, and Fig. 3.12*a* and *b* compares the cascade for a grass plot near Copenhagen on a 17-hour daylight June day (the maximum altitude of the sun being 58°) with that for a December day with 7 hours of daylight and a maximum solar elevation of 11°. Diurnal variations can also be striking, especially at high altitudes, and Fig. 3.13 compares the noon solar radiation cascade onto a bare surface at 5280 feet in the Sierra Nevada on 18th July 1963 with that at 1800 hours.

C. THE HYDROLOGICAL CASCADE

The solar energy employed in evapotranspiration (i.e. evaporation from oceans, lakes, rivers and the land surface, plus evapotranspiration from vegetation) represents, on average, some 22% of the total incident at the top of the atmosphere, but this activates a second important terrestrial

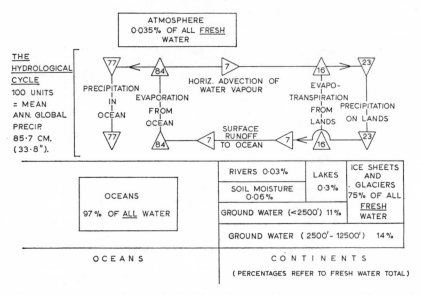

Fig. 3.14　The world hydrological cycle and terrestrial water storage. The oceanic percentage relates to *all* terrestrial water; the percentage for continental and atmospheric water to all *fresh* water. The units in the hydrological cycle are related to an assumed 100 units of mean global precipitation (77 oceanic and 23 continental). Note the relatively small atmospheric storage and the rapidity of throughput. (From More, 1967)

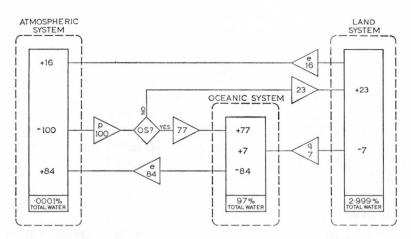

Fig. 3.15　The world hydrological cycle shown in Fig. 3.14 depicted as a canonical structure of cascading subsystems

A

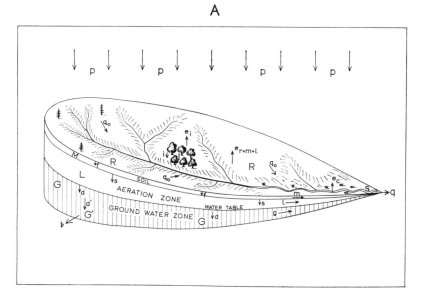

B

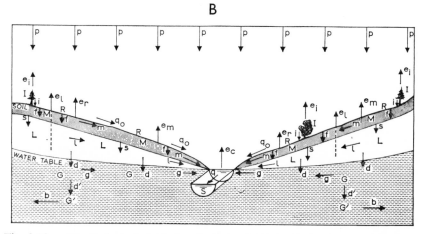

Fig. 3.16   The basin hydrological cycle depicted (A) in block diagram and (B) in generalised cross-section. (From More, 1969.) The symbols are explained in the text

cascading system, commonly termed the hydrological cycle. The generalised world hydrological cycle is illustrated in Fig. 3.14 and represented as a canonical structure in Fig. 3.15. Because the local variations in the hydrological cascade are so great, however, it is much more valuable to consider it in relation to the fundamental hydrological unit of the erosional drainage

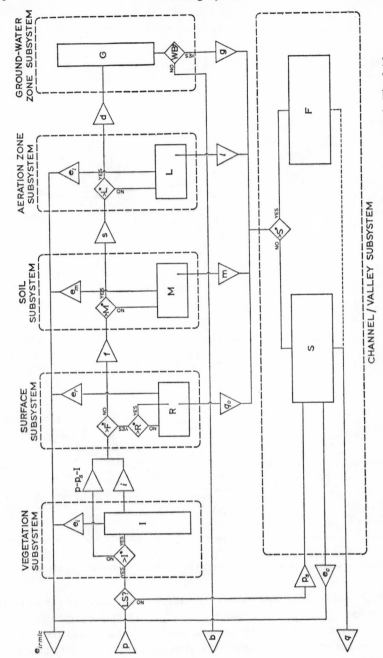

Fig. 3.17 A canonical representation of the basin hydrological cycle. The symbols are the same as in Fig. 3.16.

basin, where the collection of runoff into distinct channel flow permits its convenient measurement. The drainage basin hydrological cascade is illustrated in Fig. 3.16 and represented by a canonical structure in Fig. 3.17.

Precipitation (p) falling on the surface either falls directly onto stream and lake surfaces (p_s) or onto the land. Some of the latter is intercepted by vegetation and stored on its surface (I) and the remainder forms runoff from the vegetation. The most important regulator of the surface subsystem is the *infiltration capacity* (F^*), the maximum rate at which the surface can absorb water. Any water delivered to the surface at an intensity in excess of F^* will not infiltrate but—if it does not exceed the amount which can be retained by the surface irregularities (i.e. surface detention: R^*)—will either become part of the surface storage (R), or runoff into channels (q_0) to become streamflow. Some of the surface storage will evaporate (e_r) and will join the return flow to the atmosphere provided by evapotranspiration (e_t), evaporation from water surfaces (e_c) and evaporation of water rising from the soil and aeration zone subsystems (e_m and e_i). The infiltration below the surface (f) is operated on in the soil subsystem by the regulator of *field moisture capacity* (M^*: the maximum amount of water a given soil can hold without it draining out under gravity). The amount of f not in excess of M^* becomes soil moisture storage (M) and the remainder flows through the soil—either downwards into the aeration zone subsystem (s) or downslope sub-parallel to the surface and soil layers—as throughflow (m) to augment the streamflow. The allocation between s and m depends on the thickness of the soil layer and on the relative

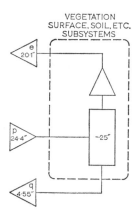
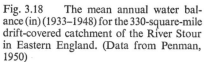

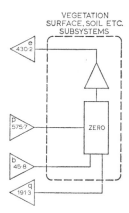

Fig. 3.18   The mean annual water balance (in) (1933–1948) for the 330-square-mile drift-covered catchment of the River Stour in Eastern England. (Data from Penman, 1950)

Fig. 3.19   The mean annual water balance (mm) (1958–1963) for a 2-square-mile catchment of flat, sandy agricultural land in southern Sweden. (Data from Högstrom, 1968)

permeability of the various soil layers, the mantle and the bedrock. For example, in areas of deep soil and steep slopes in the Appalachians as much as 16% of the precipitation from some intense rainstorms falling on moist soil becomes throughflow. Similar regulators in the aeration zone subsystem divide the flow into aeration zone storage (L), interflow (l) and deep-percolation (d) to recharge the ground-water storage (G). The slow discharge of G, with the exception of a certain amount of deep outflow to other drainage basins (b)† (though this is usually small), flows into the surface channel/valley subsystem. That not in excess of channel capacity (S^*: i.e. which can be accommodated within the stream banks) being less than bankfull discharge (q_b) becomes channel storage (S) and flows out of the basin as stream discharge (q). On the comparatively rare occasions when bankfull discharge is exceeded (on average every 1·58 years in humid temperate localities) the channel banks are overtopped and temporary floodplain storage (F) is employed.

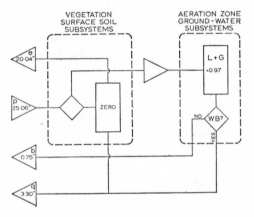

Fig. 3.20   The mean annual water budget (1961–1962) for a 42·6-acre woodland and grass watershed in Placer County, California. (Data from Lewis and Burgy, 1964). Units are inches of precipitation

The disposition of the basin hydrological cascade can be expressed by the following equation:

| consumptive use by vegetation | | changes in surface storage | changes in subsurface storage | surface runoff (q) |

$$p = \overbrace{I + e_i + e_r + e_m + e} + e_c + \overbrace{\Delta R + \Delta S + \Delta F} + \overbrace{\Delta M + \Delta L + \Delta G} + \overbrace{q_0 + m + l + g} + b$$

Of course, the measurement of all these components of the hydrological cascade in any one locality is extremely difficult, and one must usually be

† (b) may, with some subsurface conditions or after certain sequences of weather, be a net *inflow* into the ground-water storage (G).

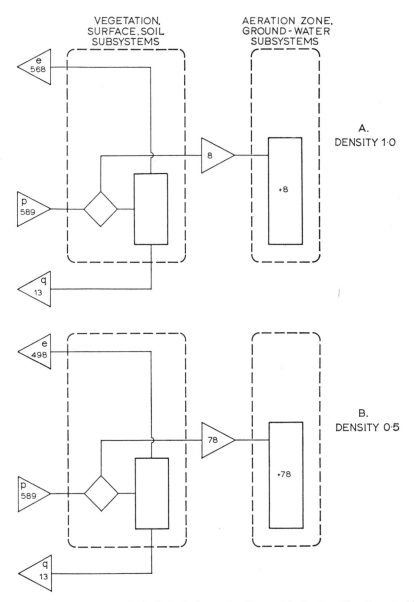

Fig. 3.21 The effect on the hydrological cascade of forest thinning in a 45 to 48-year-old oak stand in the Tellerman Experimental Forest, Voronezh Province, U.S.S.R. (*a*) shows the disposition of water before thinning, and (*b*) after a 50% reduction in crown density. (Data from Sukachev and Dylis, 1968). Units are mm of precipitation

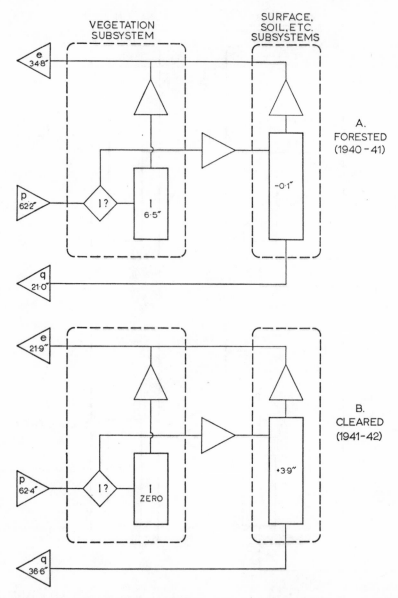

Fig. 3.22   The effect on the hydrological cascade of forest clearance from a small experimental basin in the Coweeta Experimental Forest, North Carolina. (*a*) shows the original conditions, and (*b*) those observed a year later following clearing. (Data from Kittredge, 1949). Units are inches of precipitation

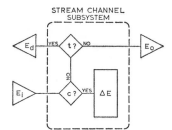

Fig. 3.23 A stream channel cascade involving energy input (E_i), dissipation due to heat losses, etc. (E_d), change in total energy (ΔE) and energy output (E_o). (Suggested by Melton, 1962)

content with much less detailed basin water budgets. The gross mean annual water balance from the River Stour catchment, a 300-square-mile drift-covered chalk basin in south-eastern England (1933–40) (Fig. 3.18), and for a 2-square-km catchment of flat sandy agricultural land in southern Sweden (1958–63) (Fig. 3.19), show the high proportion of the cascade involved in evapotranspiration in vegetated mid-latitude regions. A rather more comprehensive budget is given in Fig. 3.20 for a 42·6 acre woodland and grass catchment in Placer Co., California (mean elevation 500′) during the water year 1961–62. The considerable effects on the hydrological cascade of artificial changes in vegetation are shown by the result of halving the density of 45–48 year old experimental oak stand in Voronezh Province, U.S.S.R. (Fig. 3.21), and by the consequences of completely clearing a small basin in a North Carolina experimental forest in 1941 (Fig. 3.22).

D. OTHER CASCADES

Just as the hydrological cascade forms a small part of that of solar energy, so all the other cascading systems with which physical geographers are primarily concerned form detailed parts of these two fundamental systems. Thus it is possible to identify, for example:

1 Precipitation cascades: where moisture condenses—probably through the agency of freezing—and is then subject to a complex process of melting, evaporation, coalescence, condensation, etc., as it cascades through the subsystems of the atmosphere to the earth's surface.

2 Weathering cascades: where water plus chemical and biological constituents are cascaded through the subsystems of the various soil and weathered rock layers.

3 Vegetation cascades: where solar energy, water and minerals move from the subsystems of the atmosphere and the soil into those of plants and

animals and back again. This cascade is perhaps one of the most complicated of the terrestrial systems and obviously overlaps to a considerable extent with that of weathering. One feature of the vegetation cascade which is noteworthy is that, as the result of 'harvesting' by man, the outputs from this subsystem frequently take place in an area which is geographically remote from the region of inputs of energy and mass.

4 Debris cascades: these are based on the translation of the potential energy derived from the height of rock and soil particles above a local base level into kinetic energy and heat produced by friction. The cascade of debris and varying quantities of water through the various slope or cliff subsystems forms the link between the weathering system and those of stream channels or waves.

5 Stream channel cascades: where the flow of water, debris and solutes through the various stream-reach subsystems results in energy dissipation, largely as the result of heat losses by radiation and evaporation, friction, chemical reactions, etc. (Fig. 3.23), as potential energy is translated into kinetic energy and heat loss.

6 Valley glacier cascades: where the inputs of snowfall, potential energy of elevation and debris are cascaded from one climatic environment to another, with progressive loss of mass (through deposition, evaporation and melting) as well as the dissipation of energy as heat.

7 Wave cascades: where wave energy is dissipated across beach zone subsystems up to the edge of the swash zone, by the friction caused by transporting debris and where mass is lost by downward seepage.

Although we have isolated only nine cascading systems it should be clear that, in reality, the physical geographer is always likely to be concerned with portions of several, operating simultaneously upon one or more morphological structures. It should also be apparent that, the smaller the scale of the cascade(s) with which we are concerned, the more detailed our knowledge of the transfers of mass and energy needs to be and the more difficult it becomes to obtain this information. This latter problem must now be considered more fully.

3.3 The Description of Cascading Systems

Fundamental to all cascading systems are inputs, storages and outputs of energy and mass. We cannot proceed very far in the description of any cascade until we know at least the gross magnitude of its inputs and outputs over some period of time and, if we wish to carry out a more detailed analysis of the way the system functions, then we must know the precise location, capacity and regulators of the various stores within it. Both these aims call for measurement of mass and energy. However, while the former requires knowledge only of *total* inputs or outputs over a period, the latter demands continuous monitoring of *rates* of flow. We can, then, divide the methods of estimating the magnitude of movements of energy and mass into those which primarily *catch* and those which *gauge*.

If we now look briefly at each of the nine major cascades, we can outline the principal devices used to obtain both types of information.

A. THE SOLAR ENERGY CASCADE

Obviously, the most basic need is for an estimate of the quantity of solar energy received at the earth's surface. Having said this, it should be clear that although radiant energy is received, stored and emitted over the whole of this surface, we can only hope to measure the quantities involved at certain points. This is a fundamental difficulty and one which applies equally, as we shall see, to the estimates of precipitation, evaporation and evapotranspiration: in all cases, generalising values from points to irregular surfaces introduces errors.

The energy in the solar cascade may be measured in several different ways: Q (the direct beam radiation) is estimated by *pyrheliometers*; $Q + q$ (the diffuse radiation) by *pyranometers*; I (the long-wave radiation from the earth's surface) by *pyrgeometers*. $Q + q$ and I may be measured simultaneously by one instrument, the *pyrradiometer* and the radiation balance of any point $(Q_s - R)$ can be estimated directly by the *net radiometer*. Fig. 3.24 illustrates one of many types of pyrheliometer.

These various devices may either be read at intervals, to give the total input or output of radiation at a point over time, or may be used to record continuously.

Of these instruments, the pyranometer is perhaps the most versatile, as it may be used, with modifications, to measure Q and q separately, together with Qn (the solar intensity) and α (the albedo). For many purposes, however, the net radiometer is the most efficient instrument, as it provides an immediate value for the change in storage (positive or negative) at any point.

One of the features of the solar energy cascade of most immediate importance to the physical geographer is the variation in inputs on surfaces inclined at different angles in different directions. Fig. 3.25 illustrates the magnitude which such differences may attain even in the lower mid-latitudes and it should be clear that the difference in the intensity of direct-beam radiation

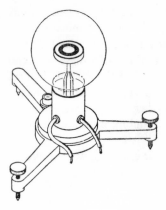

Fig. 3.24　　The Eppley pyrheliometer (180°
pyrheliometer). (After *Handbook of Meteorological Instruments*, Part I, British Meteorological
Office, 1956. From Sellers, 1965)

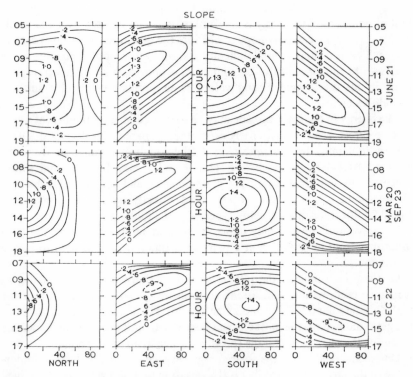

Fig. 3.25　　The average direct-beam solar radiation incident at the surface at Tucson,
Arizona (32°N), as a function of surface slope and direction, the time of day, and the time
of year. The units are langleys per minute. (From Sellers, 1965)

between north- and south-facing vertical walls will become very marked indeed as one moves polewards. Although it is apparent that the amount of energy in the solar cascade at any point will exert a direct control over the nature of the hydrological, weathering and vegetation subsystems, the difficulty of providing large numbers of measurements of even net radiation over a small area has so far proved a severe limitation to the study of this vital section of the terrestrial cascades.

B. THE HYDROLOGICAL CYCLE

At the very least the physical geographer wishes to know the precipitation (p) and stream discharge (q) of any basin. A more detailed understanding of the nature of the hydrologic cycle depends, however, on additional information concerning rainfall intensity, infiltration capacity (F^*), potential evaporation (e_c), evapotranspiration (e_i) and soil moisture ($S_M + m$). These variables differ considerably in the ease with which they can be measured and, hence, the degree of detail in which they are known. If we consider a 4th order drainage basin of perhaps 100 km^2 in part of a developed and densely-settled country such as Britain then it would, perhaps, not be too optimistic to assume the following density of recording instruments:

6 precipitation gauges;
2 streamflow gauges;
1 evaporation pan;
1 evapotranspiration measurer;
1 soil moisture measurer.

Clearly, estimations of total precipitation are likely to be more accurate than those of evapotranspiration, for example, but even so one must recall that one is generalising from 6 points to an area of 100 km^2 and that the effect of local winds and topography may produce subsystems with inputs which far exceed or fall short of the 'average' for the basin as a whole.

What sorts of instruments may be used to measure the various components of the hydrological cascade?

Precipitation

The simplest of all instruments is the standard rain gauge, which is visited once or twice a day, the catch emptied and its volume recorded. More elaborate are the recording gauges. All, however, except in the most closely-observed sites, are liable to error, the largest single source of difficulty in many localities being the measurement of snow. It is generally true that precipitation cannot, at present, be estimated more accurately than within 5% of the true value (and this figure is far too small in areas of dense vegetation, particularly in forests). This fundamental uncertainty needs to be borne clearly in mind whenever one is dealing with the description or analysis of local hydrological cycles.

Rainfall intensity, which is perhaps of even greater geomorphological importance than total precipitation, can obviously only be estimated at stations with continuous recorders.

Evaporation and evapotranspiration

How best to measure or estimate these quantities is a problem which has taxed meteorologists for many years. There are basically three approaches.

(a) The measurement of potential evaporation (e_c), either from a controlled free water surface, such as a standard evaporation pan; or by gauging inflow and outflow from a large water body, such as a lake or reservoir. An alternative method is to employ a continuously saturated porous surface, such as a Bellani or a Piche evaporimeter. All these methods are subject to considerable error and, more difficult, the relationship between the values of e_c obtained by each one is extremely complex, depending markedly upon wind speed and time of year.

(b) The measurement of potential or actual evapotranspiration (Pe_i and Ae_i): the former representing the moisture loss from a vegetated surface which is kept continuously at or above field capacity. The most successful direct measurements of these quantities have been achieved by the use of *lysimeters* (Fig. 3.26) which are large tanks filled with soil and vegetation similar to that of the surrounding area and supported on some type of weighing mechanism. Lysimeters are generally used to measure Ae_i and rather similar devices employed to estimate Pe_i are termed *soil evaporimeters*. In all cases, the location of the measuring device and its physical positioning require great care and continual supervision. This is clearly a rather expensive method of obtaining information and lysimeters are, in consequence, few and far between.

(c) The estimation of Pe_i and Ae_i by climatological methods. Taking Pe_i first, there are a great number of formulae available, but the most valuable, such as the equations of Budyko and Penman, are based upon the solar energy balance equation and involve knowledge of the air temperature, the air humidity and the available radiative energy. A less complex method has been developed by Thornthwaite, relying on an empirical relationship between mean monthly temperature and Pe_i,

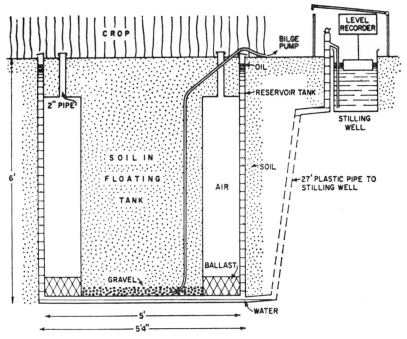

Fig. 3.26   Lysimeter installation at Hancock, Wisconsin. The observed block of soil floats in a tank of water and water losses by evapotransportation from the block are measured by recording changes of water level, instead of weighing the block. (After King. From Barry, In Chorley, 1969)

but the formula can be applied only where the air temperature is between 0 and 26·5°C. Fewer methods have been developed for the estimation of Ae_i, but the most generally useful is Thornthwaite and Mather's simplification of a suggestion by Budyko: as with the Pe_i formula, there are difficulties in applying the equation where average air temperatures drop below 0°C.

On the whole, physical geographers have tended to neglect the more precise measurements of e_c, Pe_i and Ae_i derived from instruments and to rely rather heavily upon the climatological estimates, particularly the two Thornthwaite equations. It is clear that such procedures can produce only a generalised picture of evaporation and evapotranspiration at any site.

Soil moisture
The installation of the lysimeter obviously provides detailed information about soil moisture storage changes, as well as data on evapotranspiration,

9

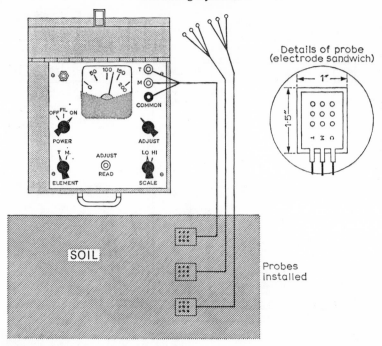

Fig. 3.27   The Soil Test soil moisture metre. At the left, an illustration of the instrument in operation, with a number of probes installed in the soil. At the right, an enlargement of the probe itself

but has the grave disadvantage of applying only to an area of horizontal ground. The physical geographer is more often interested in studying the variation in soil moisture on slopes and therefore requires a more adaptable method of measuring this feature.

The simplest procedure is to take repeated samples of soil from the same area and then determine their moisture content in the laboratory. This is, however, obviously inaccurate, since successive measurements must refer to different sections of soil.

More acceptable are the various techniques based upon the permanent or semi-permanent installation of probes of different types, the most common involving *electrical resistance units* (such as the Soil Test apparatus shown in Fig. 3.27), *gravimetric plugs* and *tensiometers*. All of these devices are small, comparatively cheap and the reading intrument readily portable. For permanent sites, the *neutron-scatter* method can also be used. In each case the device must be carefully calibrated when installed and it is desirable to check from time to time that the recording surface is still in close contact with the soil and unharmed by salts. It is generally not feasible to obtain a

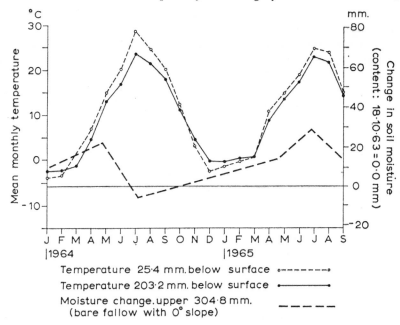

Fig. 3.28   Data on soil temperature and moisture changes—the latter measured with a neutron-scattering device—over an 18-month period at Archer Dryland Field Station, Wyoming. (Data courtesy Mr. E. Dowding, University of Wyoming. After Kennedy, 1969)

continuous record of soil moisture changes with these instruments, but a series of readings, each of which will give the net gain or loss over the intervening period. Fig. 3.28 shows the type of data which may be obtained, in this case by a neutron-scatter procedure.

Stream discharge

Measurement of the volume of the liquid portion of stream flow is a comparatively straightforward procedure and can be refined to a much higher level of accuracy than that of any other section of the hydrological cycle. It is necessary to bear this in mind when considering sophisticated methods of relating precipitation to stream runoff, as, in many cases, the latter figure is more reliable than the former.

There is a wide variety of gauging devices available, the usefulness of all of which depends upon the accurate survey of the channel cross-section at which they are employed. The simplest of all techniques is a *stage gauge*, which is simply a graduated post, against which the water level is read from time to time. More valuable are the different types of *recording gauge* which may or may not be used in conjunction with artificial cross-sections such as that provided by a weir.

Such procedures, however, give only the depth of flow, while discharge is obviously the product of depth and velocity. Measurements of the latter may be made either by hand or winch-operated *current meters*, lowered to varying depths at different positions across the channel, or by injecting *salt* or *dye solutions* of known concentration into the flow and observing the speed with which they become diluted. The latter method is particularly valuable for reconnaissance studies in rather remote areas.

Most of these methods are only successful in streams with fixed channels and fine-calibre loads, where the nature of the gauged cross-section does not vary rapidly. There are, in addition, obvious problems attached to the accurate measurement of over-bank flow.

It can be seen that to obtain information on even the inputs and outputs from the hydrological cycle in one basin is rather difficult and that any answer is, at best, an approximation for there will be substantial error in the measurement of precipitation and there may be similar, though lesser, uncertainty concerning the value for discharge (some of the precipitation, for example, may seep underground into adjacent catchments). The estimation of the quantities of water involved in the subsystems of evaporation, transpiration, throughflow, soil moisture storage and so forth is, however, even more difficult and the answers obtained will always give only a partial description of the operation of the cascade.

C. THE PRECIPITATION CASCADE

The only features of this system with which the physical geographer is generally concerned are the temperature and humidity of the air and the actual volume of precipitation received. The problems attached to estimates of the latter quantity have already been mentioned.

Measurement of air temperature and humidity is usually carried out by means of a pair of *wet and dry bulb thermometers*, which may either be mounted in a Stevenson screen, or in a *sling psychrometer* (a device resembling a large football rattle). Continuous measurements of both temperature and relative humidity may be obtained by the use of a *hygrothermograph*, although the accuracy of this instrument needs to be checked frequently against readings from a sling psychrometer.

D. THE WEATHERING CASCADE

The concept of weathering as a cascade is based upon the view that the horizons of the soil and mantle are subsystems through which heat, water, chemical solutions and clastic debris cascade, altering the size, composition, structure and location of the subsystems. Fundamental to the description of this cascade is knowledge of the volume of soil moisture, which requires that the readings of the instruments described previously be calibrated in the laboratory against the volume changes in similar blocks of soil. As the temperature of the soil exerts a fundamental influence on the speed of chemical reactions, it is also necessary to measure and, where possible, record the changes in soil temperature: this may be done almost simultaneously with soil moisture using instruments such as the Soil Test (Fig. 3.27) or *thermocouples* may be installed.

The successful operation of the weathering cascade will presumably be accompanied by the physical disintegration of the bedrock or raw debris and also by the release of minerals from their original chemical bonds. To assess the operation of this system therefore requires some measurement of the change in physical composition, mineral and organic content between the different soil and rock layers. Ideally this should be achieved by repeated recording at the same profile, but is much more often carried out by taking different samples in the same area.

The basic problem involved with measuring the operation of the weathering cascade is: how can you gauge a flow which is not contained in a definite channel without disrupting its movement and, therefore, changing the operation of the system? Clearly, if we 'catch' the water and debris percolating from one soil zone to the next, we must have inserted some sizeable piece of apparatus which, of itself, is likely to change the soil structure and therefore alter the processes in the weathering cascade. This difficulty is almost as basic to the study of cascading systems as that derived from the generalisation of point data to a large area, since it affects all flows to a greater or lesser extent.

E. THE VEGETATION CASCADE

The fundamental inputs into this system are solar radiation and precipitation, therefore we need measurements of both of these quantities. As most vegetation is layered in some way and as different amounts of radiation and precipitation may be expected to be input into each layer, it is generally necessary to have several tiers of radiometers and rain gauges, particularly if we are concerned with forest cascades. Further sources of inputs are the soil and ground water, providing many of the necessary minerals for plant growth (although some of these are acquired directly from the atmospheric cascade). In all cases, sampling of the input source is required, followed by laboratory analysis of the chemical content.

The outputs from the vegetation cascade are in several forms, the most important being heat, oxygen and carbon dioxide, water vapour, carbohydrates and other organic compounds representing 'fixed' minerals. These outputs may be measured in various ways either by 'harvesting' production from time to time—this may take the form of clean cutting small patches or of collecting the litter under plants—and determining its content in the laboratory, or by attaching *respirometers* and *calorimeters* directly to the growing plants. The latter methods provide data on *gross productivity*, whereas the former can only measure *net production* (as they make no allowance for losses due to respiration).

Unlike many of the other cascades with which the physical geographer is concerned, that of vegetation is comparatively amenable to the study of flows and storages. The most important technique in this respect is the use of *radioactive isotopes* to 'tag' inputs of water or minerals and these substances may then be monitored very accurately as they move through the different sections of the vegetation and out into the soil, ground water or animals.

Much of the study of vegetation cascades has, to date, been carried on purely by ecologists and other biologists. However, it is becoming increasingly clear that the volumes of energy and mass involved in these systems are sufficiently great to warrant attention by geomorphologists. Recent years have shown a rise in the number of interdisciplinary studies of small water-sheds, in which the role of the vegetation cascade is examined simultaneously with those of the hydrological cycle and stream-channel flow, in particular.

F. THE DEBRIS CASCADE

This system is one of the greatest interest to the geomorphologist, but presents the most severe difficulties of measurement.

Inputs into the cascade come from the solar energy, hydrological, weathering and vegetation systems, but the most important derive indirectly from the stream-channel cascade, as they are provided by the potential energy of the height of the rock or debris mass above local base level (PE_H). Estimation of PE_H has rarely been attempted, but one method which may be promising is based upon the distribution of area with height within the drainage basin: that is, upon the *hypsometric integral* (I) (see Fig. 2.18). PE_H may be calculated by combining I with the mass of the bedrock and of the regolith, both of which require knowledge of the specific gravity of the material involved and, in the latter case, additional information concerning the volume of the regolith, which may be obtained either by *augering*, or (and more accurately) by *gravimetric survey*. PE_H is at a maximum for sound bedrock at the head of a slope and at a minimum for poorly-consolidated debris at the channel margin.

Both the output from and the flows within the debris cascade are extremely difficult to measure. This is because we are concerned, by definition, with movements of semi-solid masses. The difficulties are increased, however, by the fact that movement within the debris cascade occurs in discontinuous bursts and in some systems with long *relaxation times* (for example, cliffs in semi-arid areas) the interval between bursts may be on the order of several hundred years. A variety of techniques has been used to estimate flows and outputs from debris cascades, the precise method employed depending fundamentally upon the calibre of material involved and, to a lesser degree, the volume of water contained in the flow. For large-sized material, the most economical method is to mark surface fragments and survey their positions accurately from time to time. With finer debris, the insertion of *strain gauges* may be possible, or the installation of *Young pits* or *T-pegs*. In all cases, except the surface marker technique, there are serious problems created by the installation of measuring devices, as these almost certainly distort the pattern of flow to a greater or lesser extent.

An alternative to these procedures is possible when dealing with cascades of fine-grained and relatively slow-moving debris: this is the construction of '*sediment traps*' of various types at the foot of a slope, to give the output of the debris system. Again, however, it has been found to be extremely difficult to install such traps in such a way that they do not destroy the natural pattern of movement (usually by ponding-up or accelerating soil water percolation).

G. THE STREAM-CHANNEL CASCADE

Here we are concerned with inputs and outputs of energy and mass in the forms of stream velocity, discharge, solid and suspended load volumes. Although there will be additions and subtractions from the mass of the stream along any reach without tributaries—as the result of precipitation and evaporation, bank-caving and point-bar formation, seepage into and out of the channel—it is usual to determine only the *balance* of these processes, by subtracting the outputs of the reach from the inputs.

The problems related to gauging the discharge of water have already been covered. The determination of the dissolved and suspended sediment load is slightly more difficult, but there is a variety of *suspended-sediment samplers* available (Fig. 3.29)—ranging upwards in sophistication from a bottle suspended on a string—any of which will provide a reasonably accurate estimate of these quantities. The difficulties of assessing the volume of bed load are closely akin to those involved in measuring flows in the debris cascade and, although a variety of sediment traps have been devised, it is generally recognised that there is no satisfactory solution to this problem at the present time.

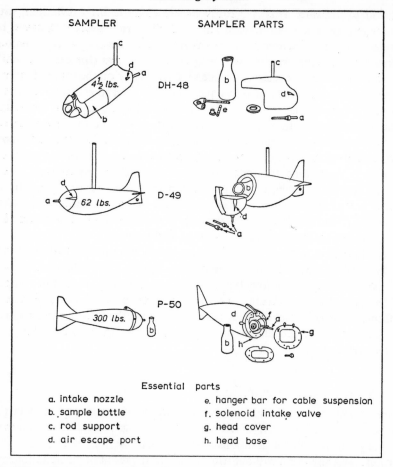

Fig. 3.29   Types of suspended-sediment sampler. (After R. E. Ottman, in United Nations/World Meteorological Office, 1962)

H. THE VALLEY-GLACIER CASCADE

The basic inputs into this cascade are of precipitation and solar energy, the major outputs water vapour, water and ice, and heat. In many respects this system shares the characteristics of the debris and the stream-channel cascades, although additional problems are created by the geographic locations of most valley glaciers.

Fig. 3.30 shows the arrangement of observation sites on one valley glacier. The *stakes* are used to measure net *accumulation* or *ablation* over the *mass balance year* (which runs from the period of minimum mass in one year, to the next): this measurement may be made either directly, or by

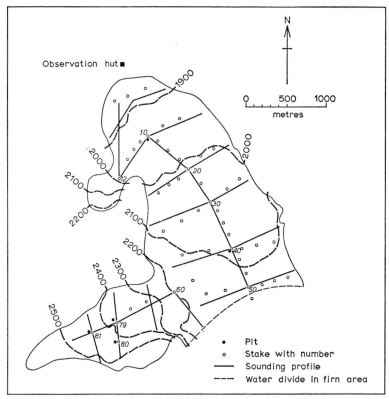

Fig. 3.30   Location of stakes, pits and sounding profiles over the period 1964–5 on the Place River Glacier, British Columbia. (After G. Østrem. From Marcus in Chorley, 1969)

photogrammetry. The *pits* are employed to obtain measures of *ice density* and the *sounding profiles* (transversed by gravimetric or *seismic* techniques) provided data on the volume of ice.

Rates of flow are far more difficult to obtain and methods employed vary between the repeated survey of surface stakes and the installation of deep cores, fitted with *strain gauges* or other devices.

As with stream flow, it has proved almost impossible to calculate the debris load of existing valley glaciers (with the exception of the surface material) and the best estimates available are based upon observations of newly exposed moraines.

I. THE WAVE CASCADE

The ultimate sources of input energy and mass into this cascade are airflow and the hydrological cascade: the importance of the former is, however, much more immediate.

As wind blows across the surface of the ocean, the three measurable characteristic features of the wave are produced: these are its *length* (*L*), *height* (*H*) and *period* (*P*). Wave velocity in the open sea may be calculated directly from observation of wave period; both wave steepness and wave energy are derived from measurements of *H*, *L* and *P* (*P* is the most readily observed). Once formed, the energy of a wave is of two sorts: potential energy produced by the elevation of the crest above still water level, and the kinetic energy derived from the movement of water particles within the wave. The relationship between these two quantities is rather complex, but it has been found that the total energy of a wave train moves half as fast as the form of the waves themselves. In contrast, the speed of transport of mass by wave trains is exceedingly slight compared to that of the train itself, except when the waves break.

As the wave train approaches the shore, energy is lost in friction, with the result that velocity declines until, in very shallow water, it becomes a function solely of water depth (*d*). The changes in wave height and steepness in shoaling water are less straightforward, but both values increase, the latter very rapidly where the ratio of *d/L* is less than 0·06. Although some energy is lost as soon as the waves come in contact with the bottom, more important is the local concentration of energy as a result of the *refraction* of wave patterns by the configuration of the coastal zone.

The measurement of the disposition of mass and energy in the shore zone becomes extremely complex, although the importance of wave period and steepness as general indicators remains. Estimation of the volume of solids transported through the cascade is also difficult, the most satisfactory solution being the accurate surveying of beach profiles between successive high tides.

3.4 The Analysis of Cascading Systems

It can be seen from this discussion that the measurement of flows within the major terrestrial cascades is by no means simple. We should, perhaps, stress once more that the major problems arise from the need to generalise the characteristics of continuous inputs or outputs from observations made at points, and from the impossibility of measuring many flows without distorting their form.

It is also clear that not only the description but the analysis of cascading systems is complex. The first step towards reducing this complexity to more manageable proportions is to represent cascades as canonical structures and a further advantage of this procedure is that it provides the basis for computer *simulation* of the way in which the systems operate. This allows a detailed investigation of the functioning of each system and enables us to predict the

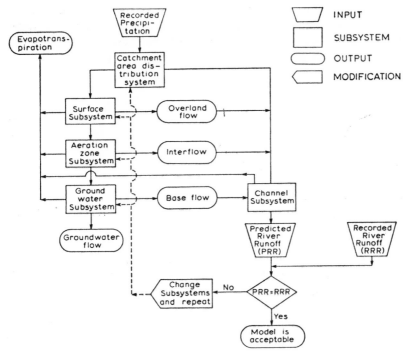

Fig. 3.31 A simple computer flow diagram describing some of the basic exchanges and storages involved in the basin hydrological cycle. (From Amorocho, 1965)

probable consequences either of changes in the natural energy inputs or of human intervention in the cascade.

For example, if we add more symbols to the drainage basin cascade, we can develop the drainage basin runoff computer flow diagram: a simplified example of such a flow diagram is given in Fig. 3.31. It is immediately clear that most of the decision regulators have been removed and replaced by subsystem processing stores. This is because most decisions regarding the relative apportionment of the cascade flows are extremely complex, while computers are only capable of making yes/no decisions. As a result, computer simulation proceeds by replacing most of the cascade regulators by mathematical equations, the solution of which enables the machine to decide how the flow should be apportioned. For example, in Fig. 3.31 one of the basic operations of the surface subsystem is to control what proportion of a given intensity of precipitation hitting the surface at a given time during a storm is likely to infiltrate. Now, infiltration capacity (F^*) at any time (t) after the start of a storm varies with the initial infiltration capacity at the start of the storm (f_0), the limiting infiltration capacity of which that particular soil is capable when completely saturated (f_c) and a constant (k) depending on the particular land slope and surface cover of the basin. During a prolonged storm the infiltration capacity of a drainage basin decreases in a predictable manner (Fig. 3.32), as expressed by the equation:

$$F^* = f_c + (f_0 - f_c)\, e^{-kft}$$

This equation is composed of *parameters* (i.e. values which are constant during any particular simulated occurrence: in this case f_0, f_c, K and e, the base of Naperian logarithms) which are given to the computer to allow it to solve the equation so that the relationship between the *variables* (i.e. those physical quantities which vary in amount during the simulation: F^* and t) will enable the machine to decide what portion of the rainfall, if any, during any period, is likely to be in excess of F^* and therefore become overland flow (or evaporate).

Much practical and theoretical work is involved both in representing each regulation decision by a mathematical expression and in calculating the values of the parameters which should be fed into the programme to represent different conditions. However, if these requirements can be met, then the computer can be programmed and the runoff cascade of the basin can be simulated by introducing a given pattern of precipitation and observing the resulting stream runoff pattern. It is possible to compare the stream runoff predicted by the computer and actual recorded runoff, by feeding in the precipitation patterns which have occurred in the past from which the resulting stream runoff patterns are already known (Fig. 3.31). Discrepancies between the two patterns then suggest how the system parameters and equations should be modified. Repeated testing and modification of the computer model may result in a programme which simulates the behaviour of an actual stream basin so closely that it can be used to predict the stream discharge pattern to be expected from almost any amount and pattern of rainfall.

The advantages of such a computer simulation system are threefold:

1 It enables one to predict the probable pattern of runoff which may be expected as the result of any given storm: this is most helpful in planning dams and other river regulation works. Fig. 3.33 shows a more complex computer simulation flow diagram—developed for such a purpose— and the runoff which the programme predicts for the Russian River in California is compared with the observed flow in Fig. 3.34.

2 It allows the prior testing of planned intervention by man in the cascade system. Thus the effects of dam construction, deforestation, etc., can be rapidly and cheaply simulated to determine whether the results of such

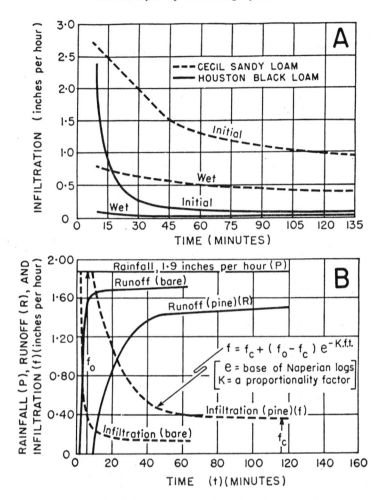

Fig. 3.32　The effects of time, soil composition, initial soil moisture conditions and vegetational cover on surface infiltration. (From More, 1967)

(a) Comparative infiltration curves for the relatively permeable Cecil and the less permeable Houston loams. In both cases the initial (dry) tests yield higher infiltration rates than the (subsequent) wet tests, with a constant applied rainfall. The curve for the wet run is not simply the last portion of the curve for the dry run, and the initial soil moisture conditions seem to influence the character of the whole infiltration curve. (After Free, Browning and Musgrave; Linsley, Kohler and Paulhus)

(b) Infiltration and surface runoff from forested land (55-year-old pines) and bare abandoned land in the Tallahatchie River basin, Mississippi, under a precipitation of 1·9 inches per hour. The infiltration curve equation is that suggested by Horton. (After Musgrave and Meinzer)

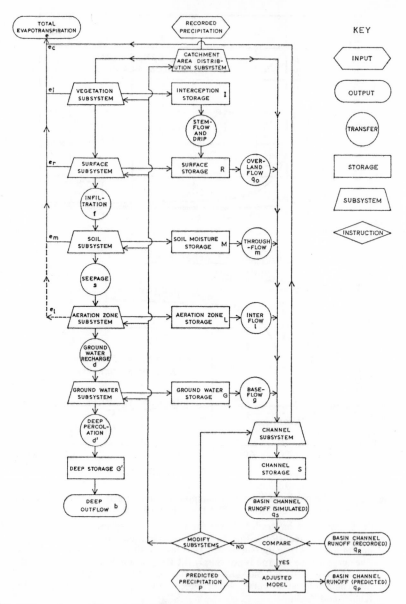

Fig. 3.33 A complex computer flow diagram describing some of the basic exchanges and storages involved in the basin hydrological cycle. The symbols are identical with those employed in Fig. 3.16. (From More, 1969)

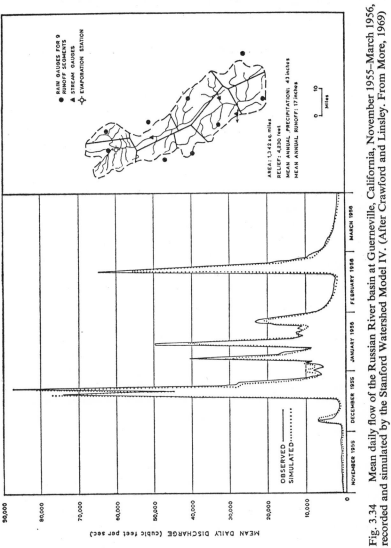

Fig. 3.34   Mean daily flow of the Russian River basin at Guerneville, California, November 1955–March 1956, recorded and simulated by the Stanford Watershed Model IV. (After Crawford and Linsley. From More, 1969)

schemes will measure up to expectation. Simulation studies are therefore frequently vital aids to the planning of resource development.

3 The construction and operation of such a complex simulation model may lead to a much deeper understanding of the nature of all aspects of the basin hydrological cascade.

However, there are also disadvantages or limitations to the use of these models:

1 It is comparatively rare to have sufficiently accurate empirical data to be sure that the parameters used are truly representative of the system. This disadvantage is, however, small where the simulation is for purely practical purposes as it is usual practice to add a 'safety margin' before applying the conclusions to real world situations.

2 Many of these simulated systems required for planning purposes achieve the correlation of measured inputs and outputs via extremely sophisticated mathematical equations which are rather loosely based upon empirical observations of the rates of change in storages in the cascade itself. This is scarcely surprising when we recall the complexity of most systems, but the physical geographer should be very cautious in transferring the conclusions from such simulations to the landscape. Just as the same morphology may result from different combinations of form and process, so the relation of one output to one input may occur via different internal flow patterns.

3 It is frequently necessary, when constructing simulation models, to hold constant the morphological variables which are important in the cascade, or at least to let them vary in a regular fashion. This necessity not only introduces a substantial element of unreality, but also may divert attention from the fact that the operation of any cascade results in the alteration of the interlocking morphological structures, and this, in turn, affects the manner in which subsequent inputs are apportioned.

Despite these limitations, the use of computer simulation models is frequently desirable and their development, testing and operation is naturally a complex matter. The type of simulation of the basin hydrological cycle described earlier is often referred to as a *white box system* (see Fig. 1.4), because its construction requires such a detailed knowledge of all aspects of the real cascade. This contrasts with the *black box system* which, in the present context, might be represented by mathematical attempts to predict stream runoff in any basin from given precipitation inputs, without any detailed consideration of how the water cascades through the system. An intermediate approach is that of the *grey box system*, in which certain limited assumptions are made regarding the internal operation of the cascade, so that outputs can be reasonably predicted from known inputs. An example of this last approach is given in Fig. 3.35, where it is desired to predict the surface runoff pattern from recorded rainfall. The only assumptions made about the internal operation of the basin cascade are: that it is possible to divide the rainfall into infiltration and rainfall excess (i.e. that flowing off the surface); that one can divide stream runoff into that emanating from subsurface baseflow and surface runoff; and that amounts of infiltration and rainfall excess are equal, respectively, to baseflow and surface runoff. Known, paired records

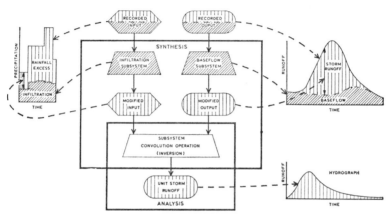

Fig. 3.35   Basic flow chart for partial system synthesis of a grey box system involving the prediction of runoff from precipitation characteristics, with a minimum knowledge of the internal operations of the system components. (After Amorocho and Hart, 1964. From More, 1969)

of storm rainfall and the resulting streamflow from the basin are fed into the programme, the mathematical equations are then invoked to operate as the subsystems and predict which parts of the input become infiltration and baseflow; these two apportionments are compared and if they are not within some predetermined degree of similarity, modifications are made to the mathematical parameters within the two subsystems (according to a predetermined plan). The whole operation is repeated until, for any one known storm, the predicted surface runoff is equal to the predicted rainfall excess. The simulation model is then further refined by testing it against a variety of past storms and their resulting runoff. The expectation is that, for the given basin, it will ultimately be possible to predict the pattern of surface runoff from almost any theoretical distribution of rainfall.

It can be seen that such a model as the last, which does not involve a detailed consideration of all aspects of the cascade system has many practical advantages, but it is necessary, when employing such a form of analysis, to keep the limitations of the procedure very clearly in mind. As a result of the difficulties in measuring inputs, storages and outputs in most natural cascading systems such grey box models are frequently used to analyse these structures.

3.5 Regulators

This extended discussion of the description and analysis of the flows through cascades has touched rather lightly upon the crucial role of the regulators within such systems. It should be stressed here that one of the most important features of the use of computer simulation models of cascades is the emphasis placed upon the key positions occupied by the various decision regulators. Obviously, the disposition of the various flows of mass and energy through cascading systems is highly influenced by the operation of variables such as the absorption capacity of the atmosphere, the albedo of the earth's surface, the infiltration capacity of the soil and soil moisture field capacity. The importance of such regulators, in terms of physical geography is, however, all the greater because of the two additional properties which they possess:

(*a*) They sometimes represent, as we have stressed, vulnerable points at which human intervention may change or modify the operation of the natural system. (We shall discuss this topic more fully in the final chapter.)

(*b*) The regulators often represent the points at which morphological and cascading systems interlock. This may be because the regulators are themselves morphological variables (e.g. amount of vegetation cover), or because they are largely controlled by a morphological variable which is embedded in a morphological system (e.g. infiltration capacity, which is primarily controlled by the geometry of soil pore spaces). Chapter 4 considers the question of the linkage of morphological and cascading systems in greater detail.

REFERENCES

Amorocho, J. and Hart, W. E. (1964), A critique of current methods in hydrologic systems investigation; *Transactions of the American Geophysical Union*, **45**, 307–321.

Amorocho, J. (1965), *Glossary on Parametric Hydrology* (Tentative), (Mimeo.)

Bormann, F. H., Likens, G. E. Fisher, D. W. and Pierce, R. S. (1968), Nutrient loss accelerated by clear-cutting of a forest ecosystem; *Science*, **159**, 882–884.

Chow, V. T. and Kareliotis, S. J. (1970), Analysis of stochastic hydrologic systems; *Water Resources Research*, **6**, 1569–1582.

Dooge, J. C. I. (1968), The hydrologic cycle as a closed system; *Bulletin of the International Association of Scientific Hydrology*, Year XIII (1), 58–68.

Gates, D. M. (1962), *Energy Exchange in the Biosphere* (Harper, New York), 151.

Geiger, R. (1965), *The Climate near the Ground*, 4th Edn. (Harvard University Press), 611.

Högström, V. (1968), Studies in the water balance of a small natural catchment area in southern Sweden; *Tellus*, **20**, 633–641.

Kirkby, M. J. (1967), Measurement and theory of soil creep; *Journal of Geology*, **75**, 359–378.

Kittredge, J. (1948), *Forest Influences* (McGraw-Hill, New York), 394.

Lee, R. (1964), Potential insolation as a topoclimatic characteristic of drainage basins; *Bulletin of the International Association of Scientific Hydrology*, Year IX (1), 27–41.

Lewis, D. C. and Burgy, R. H. (1964), Hydrologic balance for an experimental watershed; *Journal of Hydrology*, **2**, 197–212.

Melton, M. A. (1962), Methods for measuring the effect of environmental factors on channel properties; *Journal of Geophysical Research*, **67**, 1485–1490.

Miller, D. H. (1965), The heat and water budget of the earth's surface; *Advances in Geophysics*, **11**, 175–302.

More, R. J. (1967), Hydrological models and geography; In Chorley, R. J. and Haggett, P., Editors, *Models in Geography* (Methuen, London), 145–185.

More, R. J. (1969), The basin hydrological cycle; In Chorley, R. J. Editor, *Water, Earth and Man* (Methuen, London), 67–76.

Odum, E. P. (1963), *Ecology* (Holt, Rinehart and Winston, New York), 152.

Penman, H. L. (1950), The water balance of the Stour catchment area; *Journal of the Institution of Water Engineers*, **4**, 457–469.

Sellers, W. D. (1965), *Physical Climatology* (Chicago), 272.

Sharp, R. P. (1958), Malaspina glacier, Alaska; *Bulletin of the Geological Society of America*, **69**, 617–646.

Soons, J. M. and Rayner, J. N. (1968), Micro-climate and erosion processes in the Southern Alps, New Zealand; *Geografiska Annaler*, Series A, **50**, 1–15.

Sukachev, V. and Dylis, N. (1968), *Fundamentals of Forest Biogeoceanology* (Oliver and Boyd, Edinburgh), 667.

United Nations and World Meteorological Organization (1962), *Field Methods and Equipment used in Hydrology and Hydrometeorology* (United Nations, New York), 127.

Vemuri, V. and Vemuri, N. (1970), On the systems approach in hydrology; *Bulletin of the International Association of Scientific Hydrology*, Year XV (2), 17–38.

4: *Process-Response Systems*

Cause and effect, means and ends, seed and fruit, cannot be severed; for the effect already blooms in the cause, the end preexists in the means, the fruit in the seed.

EMERSON: *Essay on Composition*

4.1 The Structure of Process-Response Systems

It should by now be clear that in all the systems of interest to the physical geographer there is interlocking of morphological structures and cascades. Consequently the cascade processes can create an equilibrium state between the morphological variables and changes in the cascade system will be reflected by the changes in morphological structure, through a readjustment of the variables towards a new equilibrium relationship. Similarly, the changes in morphology may effectively alter the manner in which the cascade operates so that subsequent inputs are of different magnitude. That is to say, the inputs into such a *process-response system* at one time, *t*, consist of *both* mass and energy *and* the prior configuration of the morphologic variables. As a result of the operation of the system, energy and mass are output, *together with a new configuration of the morphological variables*. Both the form and the effect of the cascade inputs at time *t* + 1 will then be influenced by the new state of the morphological structure.

This interlocking and mutual adjustment of morphological and cascade variables is the basic feature of linkages within the process-response system, which always and inevitably take the form of feedback loops. We shall discuss the nature of these links at greater length later, but it is necessary to mention here that the most significant and diagnostic of these links involve a negative feedback loop. This exists when the configuration at time *t* + 1 is such that the subsequent disposition of energy and mass tends to return the system to the prior configuration: that is, a negative feedback loop acts in a conservative or regulatory fashion. In this way, a change in the cascade produces morphological changes which create variations in the cascading system leading to a damping-down of the effects of the original changes. This produces a new steady-state of cascade throughput (however temporarily) and a new equilibrium between the morphological variables.

In reality the structure of these negative feedback loops can be quite complex, so much so that it is often difficult to identify cause and effect clearly at every point in the loop. This difficulty is generally increased by the

126

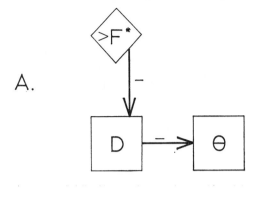

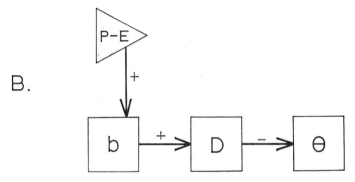

Fig. 4.1   Two simple links between cascading and morphological systems:

(*a*) The influence of a cascade regulator (infiltration capacity) upon the morphometric properties of drainage density and valley-side slope.

(*b*) The link between a cascade transfer (potential evapotranspiration) and the morphometric properties of drainage density and valley-side slope by means of the amount of surface cover (*b* = per cent bare area)

fact that most process-response systems receive simultaneous inputs of mass and energy in different forms (most obviously, water and radiation) and therefore there are synchronous adjustments in several morphological variables to the different features of the cascade inputs. The overall problems created by the interlocking of many variables—some of which may exhibit auto-correlation—frequently leads us to substitute some view based upon general functional association between variables for the more traditional cause and effect attitude.

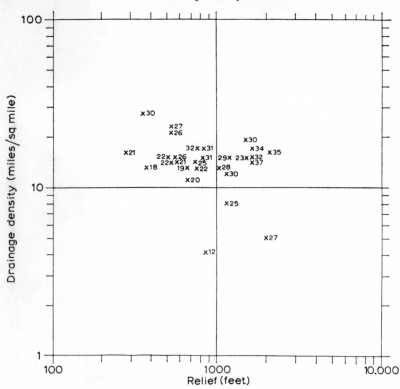

x30 = Average maximum valley-side slope angle

Fig. 4.2   The relationship between drainage density, relief and maximum valley-side slope angle, for a number of drainage basins in the western United States. Note how, for one range of relief values, slope angles rise as drainage density increases (compare Fig. 2.25), while the overall relationship (i.e. with variable relief) is much less apparent. (Data from Melton, 1957)

The process-response system embodies three important principles, which will be explored more deeply in the following chapters:

1 That the operation of the system is fundamentally controlled by the magnitude and frequency of inputs into the cascade(s).
2 That feedback loops operate to create both an equilibrium between the variables of the morphological system and a steady state of throughput in the cascade.
3 That time-directed, progressive changes in the structure and operation of the system can occur either if input changes take place or if there is some internal degradation of the system as the result of its continued operation.

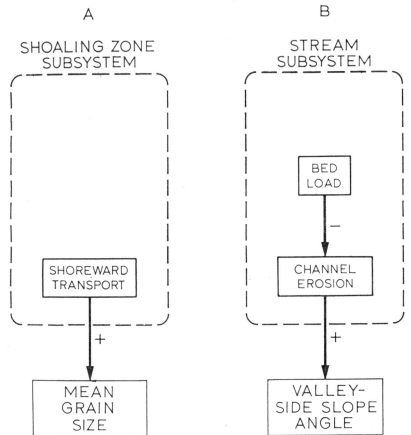

Fig. 4.3 Two simple links between cascade storages and morphological properties.
(*a*) Amount of shoreward transport on a beach and mean grain size.
(*b*) Amount of stream bed load and valley-side slope angle.

4.2 Process and Response

The links between cascading and morphological systems may be provided by variables which perform several different functions within the cascade:

1 Most important are the regulators. For example, it has been found that—other things being equal—a high infiltration capacity (F^*) results in low surface runoff, leading to a low drainage density (D) and a high valley-side slope angle (θ) (Fig. 4.1a). (It should, however, be stressed that this relationship is highly scale-dependent: that is, it applies in this form only when substantial differences in relief and climate are included in the analysis. Within one area of uniform relief and climate it is clear (see Fig. 2.25) that high infiltration capacity and low drainage density will

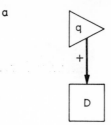

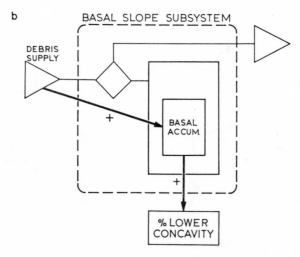

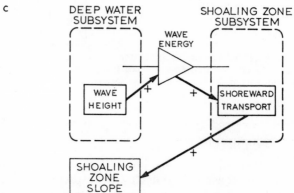

Fig. 4.4   Examples of links between cascade transfers and morphological properties.

(*a*) Basin overland flow and drainage density.
(*b*) Debris supply to the basal slope and percentage lower concavity.
(*c*) Wave energy and shoaling zone slope.

BASIN SUBSYSTEM

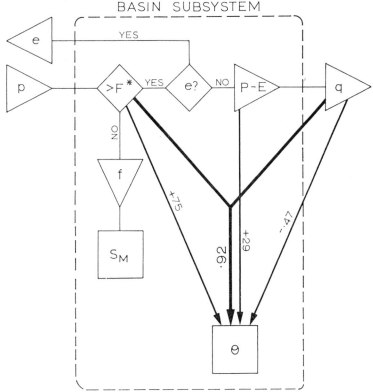

Fig. 4.5   A process-response system involving multiple correlations (*R*)
relating to valley-side slope angle. (Data from Melton, 1957)

produce *low* valley side angles. Fig. 4.2, illustrates the variable nature of
this relationship, depending upon the degree of variation in relief).

2 A second type of control of morphometric properties is that which may be
exerted by storages within cascade subsystems. This may occur in a fairly
straightforward manner as, for example, when increased shoreward
transport of material in the shoaling-wave zone produces an increase in
the mean grain size of the particles concerned (Fig. 4.3*a*). In other cases
one must introduce intervening links in order to understand the relation
between the process and the response, as when an increase in stream bed
load causes a decrease in vertical channel erosion and, ultimately, a
decrease in valley-side slope angle (Fig. 4.3*b*).

3 A third type of linkage is that provided by the transfer of energy or mass
between subsystems of the cascade. This control may be simple, as when an
increase in runoff intensity (*q*) tends to increase drainage density (Fig.
4.4*a*), or it may occur through the agency of a subsystem storage as, for

BASIN SUBSYSTEM

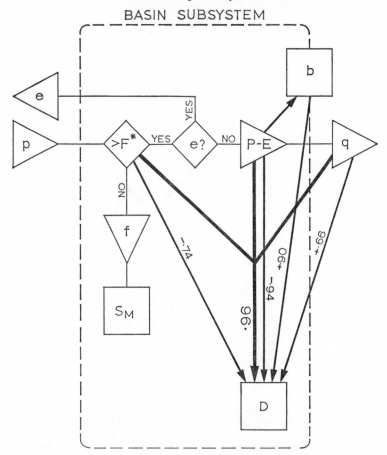

Fig. 4.6   A process-response system involving multiple correlations (*R*) relating to drainage density. (Data from Melton, 1957) (b = per cent bare area)

example, when an increase in debris reaching the foot of a slope results in basal accumulation and, in turn, an increase in the percentage of the slope profile comprising the lower concavity (Fig. 4.4*b*). From the geomorphologist's point of view such a chain of linkages is much easier to study when the transfer is itself largely controlled by some variable which is straightforward to measure; for example, the way in which wave height affects wave energy (Fig. 4.4*c*). A similar relationship, illustrating the importance of a regulator, is that between high potential evapotranspiration ($P - E$), large percentages of bare surface (*b*) and high drainage density (Fig. 4.1*b*).

4 Finally there are, of course, those cascade elements which are shared by morphological systems. Infiltration capacity is probably the most important

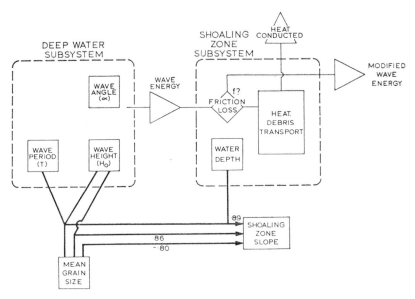

Fig. 4.7   A process-response system involving multiple correlations relating to the slope of the shoaling zone on a Lake Michigan shore. (Data from Harrison and Krumbein, 1964)

of these, functioning as both a cascade regulator and a morphological variable.

Although it is often convenient to examine the structure of process-response systems in terms of simple links alone, it is obvious that multiple correlations occur. For example, an analysis of more than 80 drainage basins in the western United States showed that, despite other significant correlations, some 83·6% of the observed variation in maximum valley-side slope angle (θ) could be attributed to the observed variation in infiltration capacity (F^*) and runoff intensity (q) (i.e. $R_{\theta \cdot F^* q} = 0.92$) (Fig. 4.5). Similarly the same two independent variables plus potential evapo-transpiration ($P-E$) accounted for 92·2% of the variation in drainage density (D) between these basins (i.e. $R_{D \cdot F^* q P-E} = 0.96$) (Fig. 4.6). In another study, wave height and mean grain size were found to account for 74·1% of the observed variation in the slope of the shoaling zone on a Lake Michigan shore (grain size individually contributing more than 63%) and the addition of wave period and water depth improved the explanation to over 78% (Fig. 4.7).

One final factor whose influence upon process-response linkages needs to be considered is the relaxation time involved, as this controls both the time required for form to respond to changing process and, by way of the feedback loop, that necessary for a new equilibrium to be achieved throughout the process-response system. This topic will be treated in Chapter 6, all that need to be said here is that relaxation time is highly variable from one system to another: for example, changes involving wave height and shoaling zone slope may be measured in minutes; those involving surface runoff and drainage density on weak clays, in months; and those involving changes in available stream bed load and valley-side slope angles, in hundreds or even thousands of years.

4.3 The Feedback Loop

Positive feedback (i.e. self-reinforcing) loops are quite common in nature but, for obvious reasons, their structure has a built-in self-destructive element which limits the time for which they can operate. For example: the positive

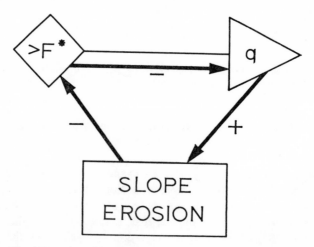

Fig. 4.8   A positive-feedback loop in an erosional drainage basin system. (After Melton, 1958)

feedback loop where a decrease in infiltration capacity (F^*) causes an increase in surface runoff (q), which causes an increase in slope erosion, leading in turn to the removal of the more permeable upper layers of the soil—and to a further decrease in F^* (Fig. 4.8) can only continue to operate as long as erosion progressively decreases the infiltration capacity. Very soon the link between slope erosion and F^* vanishes and the feedback loop is destroyed, at least for the time being. Unlike short-term fluvial processes, some glacial erosional processes exhibit strong local positive feedback which has been suggested as a reason for the great variety and apparent disorder of glacial features, compared with fluvial ones. This positive feedback can be illustrated by the process of glacial overdeepening, when a local deepening of the bed of a valley glacier produces a concave profile leading to increasingly compressive ice flow, this in turn gives a greater tendency for basal eroded material to be carried up into the ice, leading to an increase of concavity and erosion. This sequence is clearly limited in time by increased debris clogging of the ice, but while it is occurring the positive feedback may be further assisted by the increased local erosion associated with ice thickening and the tendency for break-up of the glacier bed by pressure-release jointing as glacial erosion proceeds deeper and deeper.

Negative feedback loops, on the other hand, are much more common and vital to the structure of process-response systems, where they serve something of the same purpose as governors on mechanical engines. The operation of these loops brings forces into play which tend to oppose continued change within the system caused by some variation of cascade input. This self-regulation (which was once termed *grade* when applied to stream channel processes) operates so that a new equilibrium is established after energy changes within the system. For example: an increase in basal stream erosion tends to steepen the angle of the associated valley-side slope, accelerating the rate at which debris is supplied to the channel and ultimately inhibiting the basal erosion. In this way the change in a morphological variable often exerts a regulating effect on part of the linked cascade to bring about a new condition of equilibrium. In this case the new system state might have steeper, but effectively stable, valley-side slope angles; it being, in fact, difficult to generalise about the form of the major morphological variable in such a situation, since the relationship between the values of θ at the time of undercutting and after the inhibition of basal erosion will vary according to the relaxation time of the system. Fig. 4.9 illustrates three negative feedback loops. Fig. 4.9a and b are extensions of the simple example described above. Fig. 4.9c shows two shore negative feedback loops. In one, the inverse control exercised by the slope of the shoaling zone over wave height is the key link; and in the second the increase in beach roughness inhibiting shoreward transport performs a similar regulatory function.

It is most important to remember that the situation in nature is complicated both by the fact that feedback loops interact upon each other in complex process-response systems and also that the loop links represent correlation coefficients of less than unity. That is to say, *the controls within each loop are only partial and other variables are involved.* For example, slope erosion is only one of the factors controlling stream bed load (Fig. 4.9a) and debris roughness only one of the controls over shoreward transport (Fig. 4.9c). In other words, the effectiveness and speed of self-regulation in process-response systems depend jointly upon: the strength of the individual links in

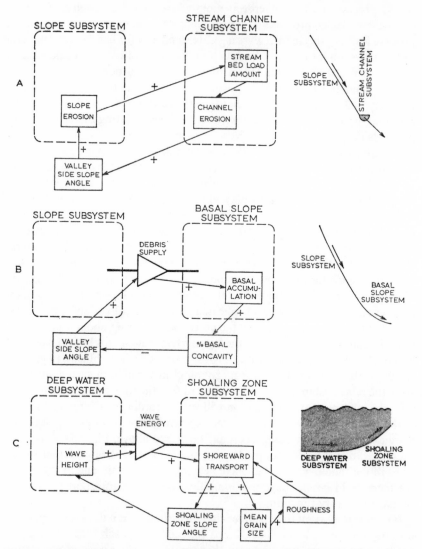

Fig. 4.9   Three examples of negative-feedback loops connecting cascading systems to morphological systems to form characteristic process-response systems.
(*a*) and (*b*) are elaborations of the examples already depicted in Figs. 4.3(*b*) and 4.4(*b*), respectively.
(*c*) Shows two suggested feedback loops in a coastal system, elaborating the relationships shown in Fig. 4.4(*c*).

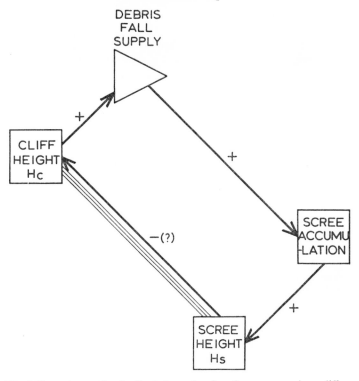

Fig. 4.10   A negative-feedback loop showing, for any one given cliff-scree system, the relation between cliff height and scree dimensions. (Data from Towler, 1969)

the feedback loops; the simplicity of organization of the system; and the various relaxation times involved.

Another feature of some feedback loops in process-response systems is that it is not always possible to deduce directly the way in which they will operate in any one case, by the analysis of a series of samples from different locations. For example, Fig. 4.10 shows a series of high positive correlations between certain morphological properties of cliffs and the form of scree slopes on the Skiddaw Slates of the English Lake District. It is clear that there is a general tendency to find high screes associated with high cliffs but, *for any given cliff*, the higher the scree extends, the more of the cliff face is obscured and the less the supply of debris. Theoretically, when the debris supply becomes reduced so that it is equal to the rate of debris creep to the base of the scree slope, an equilibrium state will obtain. A rather similar example is shown in Fig. 4.11, where the relationship between profile height and maximum angle of valley-side slope can be seen to reverse between the local and regional scale for two sets of coombes on the Chalk of southern England. Clearly, on purely geometrical grounds there is an increased probability of high slope angles as valley depth increases; within an area of roughly uniform relief, however, high slopes very often tend to be relatively long and are therefore less stable than comparatively low, short slopes. This is because

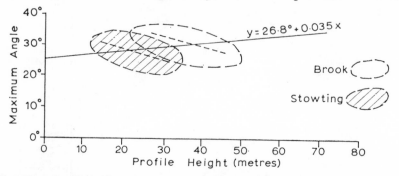

Fig. 4.11   An illustration of the manner in which scale may effect the direction of trends. When profile height (X) and maximum valley-side slope angle (Y) are correlated *within* small areas of chalk escarpment valleys, a negative association results. When the *regional* trend is computed, however, by combining the two samples, the relationship appears to be positive. (From Kennedy, 1969)

the stability of any slope section depends not only upon its angle but also upon the weight of material lying up-slope, so that long, high slopes will be likely to suffer a comparative reduction in angle.

A further difficulty is that, if only parts of process-response systems are considered, the vital negative feedback loop may be missing.

4.4 The Types of Process-Response Systems

The two preceding chapters have emphasised the ubiquity of both morphological and cascading systems at the face of the earth. It follows that both the number and the range in magnitude of process-response systems of interest to the physical geographer are vast. Rather than working through the probable linkages between the basic morphological structures described in Chapter 2 and the cascades outlined in Chapter 3, it seems more valuable, at this point, to consider the different types of process-response systems in a rather more abstract light. One may, in fact, define four classes of process-response system:

1 Those best represented as points, where the cascade elements are almost purely vertical (e.g. plants; soil sections).
2 Those best represented as lines, where the cascade has both a gravitational and a horizontal component (e.g. slope profiles; streams).

3 Those represented as essentially uniform surfaces, where the horizontal components of the cascade dominate (e.g. interfluves).

4 Those which can only be characterised as irregular surfaces, where the vertical and horizontal components of the cascade assume varying importance at different points (e.g. the atmospheric circulation; the land surface).

In reality, most process response systems in physical geography are of the fourth type, but it is frequently convenient to analyse their components as if they were of the other forms. What is perhaps most important about such a classification is that it emphasises the need to vary the methods of investigation according to the form of the process-response system.

We should now look, briefly, at some examples of these four categories.

A. POINT SYSTEMS

The most obvious of these are individual plants. By combining detailed measurements of solar radiation, precipitation and nutrient uptake with morphologic measurements of different features of the growth form, it is possible to arrive at an extremely detailed picture of the process-response system involved. However, such studies are rarely carried out by physical geographers and are probably at too small an areal scale to be of much direct relevance.

Other studies of point systems of geomorphological importance are rather more artificial in that there is selection of one point (or column of very small diameter) which is then considered as representative of a surface. Investigations of this type have been carried out on the nature of rain splash erosion, where the principal cascade elements involved are the terminal velocity of rain drops (dependent, in turn, upon the morphological variable of drop diameter) and rainfall intensity; and the major morphological variables are the grain-size and cohesion of surface particles. The output from this process-response system is the distance to which particles are moved and the form of surface depression produced. The latter feature will, of course, influence the success of subsequent splash erosion, unless pits are infilled between rainstorms by surface wash or creep. Such a point system is essentially simple and responds best to analysis by the techniques of statistical mechanics, in that it is possible to develop equations which relate the input variables (terminal velocity; rain intensity; grain size; cohesion) to the output (distance of ejection; depth and diameter of splash-pit). It is also possible to extend such an analysis to the case of a point upon a sloping surface by introducing a further morphological variable—ground slope—and here one is clearly dealing with a system in many ways transitional between types 1 and 2.

A more complicated point system and one which is more difficult to study because of the arbitrary delimitation of the system, is that comprising a vertical column of soil. Clearly the major cascade inputs into such a system are vertical (radiation; precipitation; weathered bedrock; plant and animal litter) but there may well be a substantial horizontal component as well (lateral advection of heat; soil or ground water circulation carrying with it nutrients; mixing of soil by earthworms, etc.). Nevertheless, in many cases one does wish to consider soil formation in terms of the operation of essentially vertical cascades upon essentially horizontal layers of regolith and organic debris, each of the layers representing a subsystem of the total structure. Analysis of such point systems has, in general, been rather qualitative and theoretical. A common emphasis is upon the ideal soil column, representing some particular combination of climate and parent material and considering the operation of the process-response system over a long period of time. This will produce, as the major output, a 'mature' soil profile. As with the study of splash erosion, the point selected for analysis may often be located upon a slope and consequently the sub-horizontal components of mass and energy flows will assume greater importance.

By and large the selection and analysis of point systems in physical geography has not yet proceeded very far, but it is arguable that there may be definite advantages in this approach. The chief drawback, of course, lies in extending the conclusions drawn from one such system to the behaviour of the entire surface on which the point lies. There are two reasons for this difficulty each of which applies with rather different force in different systems: first, that the point selected may not be truly representative of the surface as a whole; second, that the true functioning of the surface will involve interaction between the infinity of point systems of which it is composed and this interaction may considerably distort the operation at each point. Despite these problems, analysis of point systems does provide an insight into the fundamental characteristics of many process-response systems.

B. LINE SYSTEMS

The process-response systems that may be considered essentially as lines are much more familiar to the physical geographer. They vary, as a class, in two major respects: first, in the relative strength of the horizontal and gravitational components of flows (compare, for example, the profile of a cliff and scree system, with a river in the floodplain stage); second, in the degree to which the line represents a true channel of activity (the same comparison may be made here). By and large these two variations are linked in that the occurrence of true line systems—as opposed to artificial 'profiles'—becomes more common as the strength of the horizontal component of flows increases.

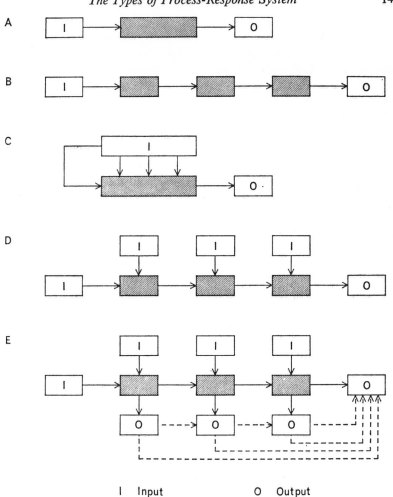

Fig. 4.12   Illustrations of the different ways of viewing input, throughput and output.
See the text for a detailed explanation

A rather problematic feature of line systems is the variation in the
location of inputs and outputs—whether real, or as a result of the measurement techniques involved—and the major categories are summarised in
Fig. 4.12.

Fig. 4.12*a* indicates the simplest case, where all flows enter the system at
one point in time and space and leave it at another: such a system must be
extremely rare—if not non-existent—in nature, but it is frequently convenient to assume that it exists as, for example, when examining the effect

of discharges of particular magnitudes upon stream channel form. One natural system which does, perhaps, approach this ideal is that of cliff and scree systems in semi-arid areas, where input occurs very infrequently in the form of huge block falls, and output is ultimately in the form of finely-comminuted debris to a stream channel. In this case, however, the output is high discontinuous.

Fig. 4.12*b* illustrates a rather more complex case, where, although input is still *en bloc*, the transfer between subsystems can be differentiated. Again, this is an unrealistic case, although frequently employed for analysis. One might, for example, envisage the passage of a given quantity of precipitation through successive soil layers until finally output as ground water to a stream channel, or the movement of the debris created by a block-fall over the various slope sections, to a channel.

The line system shown in Fig. 4.12*c* is slightly more realistic, in that inputs occur not only 'upstream' but also throughout the length of the system: output, however, is still envisaged as occurring at a point. For example, it may be moderately accurate to view the behaviour of a scree slope thinly covered by debris in this way, as it receives material both from a cliff and by weathering of the underlying bedrock, but loses it only to the adjacent thalweg. The behaviour of a stream channel section where competence and capacity are nowhere exceeded is perhaps similar, since debris should be acquired throughout the reach.

With the system illustrated in Fig. 4.12*d*, we approach reality still more closely, in that both inputs and throughputs are seen to be composed of discrete, though related, components. If we consider a complex slope profile with several sections of different angle and consequently variable soil depth, then the inputs of debris either from up-slope or from weathering of bedrock will vary, as will the amounts of material transferred between the sections.

It is, however, only with systems such as that shown in Fig. 4.12*e* that we enter the real world. Here inputs, throughputs and outputs are all seen to be made up of discrete units. Such a structure applies to most line systems and certainly to all those in which precipitation is important, as this will not occur uniformly throughout the system nor will output by evapotranspiration, seepage and runoff be as a continuous stream. If we consider any substantial length of stream channel between two junctions then it is clear that the amount of discharge received by each section of the reach will be that input at the head of the stretch, minus infiltration into bed and banks and evaporation, plus rainfall on the stream surface and seepage from ground water. Similarly, the debris and solute load will suffer gains and losses in each section of the channel. Although, as outlined above, it is common practice to adopt the 'black-box' approach to the analysis of stream channels, it is

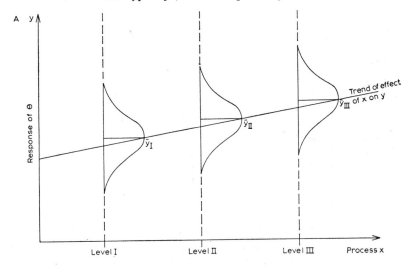

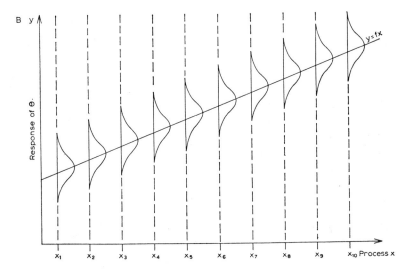

Fig. 4.13   A comparison of the techniques employed by (*a*) analysis of variance and (*b*) simple regression analysis

most important for the physical geographer to appreciate the real, discontinuous mode of operation of such line process-response systems.

This brief description of linear systems has pointed to one major problem in their analysis, namely that involving the lag-time of the system, which means that it is often much easier to identify an input event than to isolate

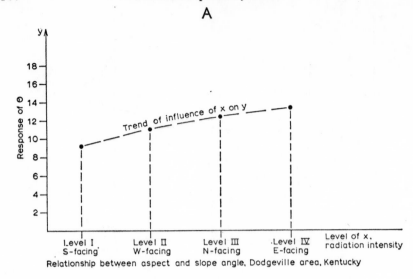

Relationship between aspect and slope angle, Dodgeville area, Kentucky

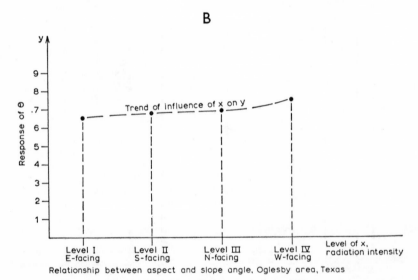

Relationship between aspect and slope angle, Oglesby area, Texas

Fig. 4.14   Illustrations of the way in which 'trends' may emerge as the result of analysis of variance and also of the fact that the 'ranking' of levels of the controlling variable (radiation intensity) can only be done with reference to the magnitude of each level's effect on Y (slope angle).

(a) The ranking for the Dodgeville area of the Eden Shale Belt, Kentucky.
(b) The ranking for the Oglesby area of Texas and slopes cut on the Edwards and Comanche Peak limestones.
(Data from Kennedy, 1969)

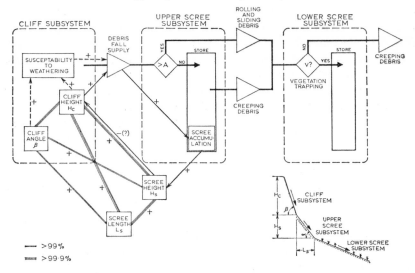

Fig. 4.15   A cliff-scree process-response system developed with general reference to the English Lake District. (Data from Towler, 1969)

the corresponding output exactly. Although such a difficulty is particularly marked in the case of semi-arid slope systems where the storages are large and the throughputs slow, it is present to a lesser extent in many linear systems and is, of course, frequently the cause of changes in the morphological variables, most particularly the angle of the slope of the 'channel' concerned.

In general, the analysis of line systems has proceeded along two main paths, by correlation and regression, and by analysis of variance. In both cases it is common practice to employ the morphologic feature of ground slope (θ) as the dependent variable (Y), even though it is realised that all linear systems pose the classic chicken-and-egg dilemma: does the process produce the slope, or does the slope govern the process? The choice between regression or variance analysis is frequently dependent upon the perception of—or ability to measure—inputs into the system. Where these may best be viewed as discontinuous, or as differing qualitatively, then variance analysis is appropriate (Fig. 4.13a); where inputs vary continuously in magnitude regression analysis is frequently preferred (Fig. 4.13b). It is important to realise the difference between these two approaches. With analysis of variance the ranking of the levels of process X is often done *purely* with respect to the magnitude of their effect on Y, and occurs after the study has been completed. For example, if one wishes to consider the relationship between variation in magnitude of solar radiation inputs, as summarised by slope aspect

(X) and profile angle (Y), it would be impossible to predict the nature of the relationship beforehand, although one might ultimately establish a trend (Fig. 4.14 illustrates this for two different areas). With regression and correlation analysis, on the other hand, the magnitude of X *must* be calculable quite independently of its effect on Y; as, for example, discharge, which may be measured on the ordinal scale irrespective of the degree of effect which each discharge may have upon channel gradient.

Clearly most studies of valley-side slope profiles and of streams (whether of water or ice) can be thought of as analyses of line process-response systems. The chief difference between the two classes is in the comparative ease with which the cascade and morphological variables can be measured. To some extent it is true to say that most slope studies up to now have emphasised the morphological features and considered cascade inputs and outputs only cursorily. On the other hand, work on rivers and glaciers frequently emphasises the variations in flow and reduces bed and bank morphology to invariable parameters in the system. Practical considerations may make this separation of process and response in the study of line systems almost inevitable, but it is necessary to bear in mind the fact it is a purely artificial division. Fig. 4.15 illustrates, for the cliff and scree systems previously discussed, an attempt to relate both morphological and cascade variables.

C. PLANE-SURFACE SYSTEMS

The clearest example of this class of response systems is given by a cultivated field, where the variable inputs of solar radiation, moisture and nutrients over the surface react with the intrinsic differences in the genetic composition of each plant to produce a non-uniform height of mature individuals. In some cases the surfaces of interest to the physical geographer are similarly characterised by a 'response' of height: as, for example, the elevation of a beach surface after each successive input of wave energy and debris; or the height of a water table after periods of differing precipitation inputs; or the depth of soil cover on an interfluve. More often however, the 'response' variable has no vertical dimensions: the grain size of beach material; the amount of moisture in the surface soil cover of an interfluve area. It should be apparent that such plane surfaces can be regarded as an aggregate of point systems and that, in fact, they are generally analysed by sampling a number of such points. As with line systems, there are two major methods of analysis available: those involving the fitting of trend surfaces and based upon the methods of correlation and regression analysis; and those based upon the analysis of variance.

The method of fitting trend surfaces has already been outlined in Chapter 2. Whilst it is in some ways eminently suited to the description of plane

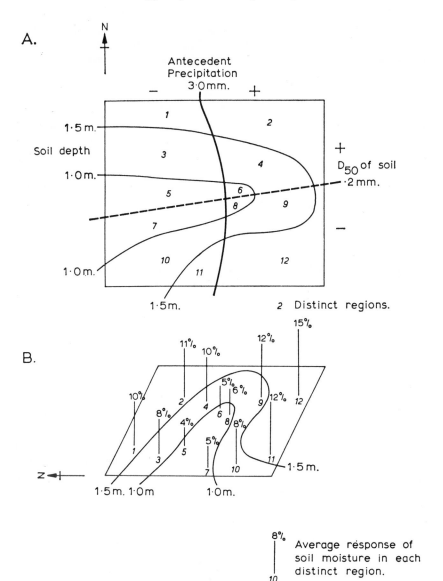

Fig. 4.16   An illustration of the construction of a process-response surface via analysis of variance.

(*a*) Separation of 12 regions which differ with respect to the 'intensity' of three different controlling factors.

(*b*) The average 'response' of soil moisture in each region.

process-response surfaces—for example, it provides an immediate visual impression of the 'intensity' of response—it does have certain major drawbacks. First of these is the fact that the 'independent' variables are always the X and Y co-ordinates of each point sampled and not the variations in cascade inputs themselves, although in certain cases—a beach, for example—location must be directly linked with volume and nature of input. Second, there is the problem of locating sampling points: these ought, from the statistical point of view, to be random, but it is far easier to fit a surface to a systematic grid. Third, and most important, there is bound to be autocorrelation between neighbouring sampling points, which leads to inaccuracy in the interpretation of the 'fit' of the surface.

Analysis of variance, on the other hand, is a technique which was devised expressly for evaluating areal process-response systems, in that it was developed primarily by agricultural statisticians. It has the great merit that it can be used to assess not only the indirect variation of inputs associated with location, but also that arising from the differing intensity of cascade features. Fig. 4.16 demonstrates this possibility, for an imaginary investigation of the response of soil moisture (S_M) in a flat interfluve area to: the amount of precipitation received in the last storm (>3mm in the east of the area; <3mm to the west); the average depth of the soil cover (<1m; 1–1.5 m; >1.5 m); and the median grain size of the soil mantle (D_{50}) (>0.2 mm; <0.2 mm). It can be seen that there are $2 \times 3 \times 2 = 12$ possible combinations of the various intensities of morphological and cascade inputs (<3 mm rainfall, >1.5 m soil depth, >0.2 mm D_{50}; <3 mm rainfall, >1.5 m soil depth, <0.2 mm D_{50}; etc.). The twelve different areas which correspond to these twelve different combinations of input are indicated on Fig. 4.16a. In order to assess the response of soil moisture to the joint effect of these three controls we need to locate an equal number of random samples—preferably in a randomly-located quadrat of uniform size—within each area. The average value of S_M might be found to vary in the manner illustrated in Fig. 4.16b, in which case the analysis of variance might show a significant *interaction* between all three inputs, but with a marked negative influence exerted by soil depths of less than 1.0 m on soil moisture.

The chief advantage, then, of the use of this type of *factorial experiment* and analysis of variance is that it enables us to evaluate the response of any feature to simultaneous variations in the input morphological and cascade variables. Moreover, and most important, it allows us to identify rather precisely the unusually high or low responses and relate them directly to a particular combination of input factors: in the example given, it seems that rather heavy antecedent rain (>3 mm) has the most lasting influence on soil moisture (15%) where the soil is deep (>1.5 m) and fine-grained (D_{50} <0.2 mm). The main disadvantage of factorial analysis of variance is that there must be as many areas or classes of output identifiable as there are possible combinations of the intensities or levels of input and, with large numbers of factors involved, it is often impossible to fulfil this requirement. There are, however, ways round the difficulty. If only one type of area is absent, then *missing plot* techniques can be employed; in other cases one can combine regression techniques with the analysis of variance, in *covariance analysis*. A further apparent drawback is that the analysis of variance does not result in a contour map of the response surface: however, as this avoids giving an inaccurately precise impression of the intensity of response, it is probably an advantage!

D. IRREGULAR-SURFACE SYSTEMS

As we have already said, these represent far and away the most ubiquitous process-response systems of interest to the physical geographer: types A, B and C, can, in fact, be thought of as limiting and rather unreal cases of this general form. A major aim of physical geography can be thought of as the explanation of the varying responses of the land surface to the differential inputs of rock type and structure, climate and history: or, to put it in W. M. Davis' words, structure, process and stage.

Clearly, this type of process-response system is at a much larger scale than any of the others. This means that we are forced to be highly selective in the degree of 'explanation' which we accept. For example, if we wish to understand the 'response' of slope angles over southeastern England to variations in structure, process and stage, we must ignore a great deal of local 'noise' which we would need to 'explain' if we were considering line process-response systems in one or two valleys within the area. Ideally, of course, one would like to have such an intensive knowledge of point, line and plane-surface systems that they could be built into a comprehensive model of the irregular surface. At present, however, we find it very difficult to make the links between the local and regional scales and consequently many 'explanations' of irregular surface systems are still at rather a crude and generalised level. We should perhaps stress that this is not necessarily disadvantageous: very often as we have seen, the effect of different inputs really does vary according to the scale considered and it is often important to try and separate 'the wood from the trees'. However, as such general explanations frequently do not hold at a more local scale it is equally necessary to avoid making incorrect assumptions about the behaviour of one stream or slope on the basis of large-scale areal trends.

There are two major groups of techniques which have been used to analyse the response of irregular surfaces: trend surface analysis and space filtering. (It is also quite possible to extend the factorial analysis of variance techniques to such systems, in much the same way as for plane surfaces.)

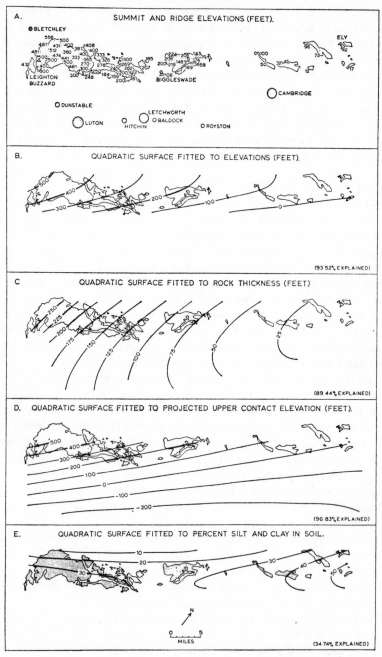

Fig. 4.17

Trend surface analysis has been applied rather rarely to irregular surfaces. Fig. 4.17, however, illustrates one such study, where the elevation of the Lower Greensand ridge in southern England (Fig. 4.17a) has first been generalised as a quadratic trend surface (Fig. 4.17b) and then viewed as a response to a group of spatially variable controls. The three such controls selected were: the thickness of the formation (Fig. 4.17c); the projection of its upper contact (Fig. 4.17d); and the coarseness of its constituent sands (Fig. 4.17e). All three controls were generalised by quadratic trend surfaces and those for the first two found to give a very high degree of fit. Finally 50 points were selected and the values of the dependent and three independent variables calculated for each. These data were used to run a multiple correlation analysis, which showed that: 98·9% of the observed variation in height along the ridge could be accounted for by differences in thickness alone; variations in the projected elevation of the upper contact added a further and significant 0·6%; and sand size variations provided a smaller but also significant contribution.

A rather similar study, of a system in which the vertical elements of the cascade are less pronounced is shown in Fig. 4.18. This represents a study of the distribution of median diameters of sand on a Lake Michigan beach. The linear surface (Fig. 4.18c) indicates a tendency for the sand to become finer along the beach in a southeasterly direction, the quadratic surface shows a pronounced decrease in coarseness up the beach from the plunge point of the breakers (Fig. 4.18d), and the residuals from the quadratic trend (Fig. 4.18e) show a further pattern of facies changes parallel to the shore. These tendencies can be respectively rationalised in terms of the angle of wave attack from the dominantly north-easterly fetch; the deposition of coarser sands in the plunge zone and the carrying of progressively finer particles up the beach by the swash; and by possible changes in the plunge point under different storm wave conditions.

These studies illustrate one of the major drawbacks of trend surfaces: that one can *only* fit surfaces to the distribution of *one* variable at a time (whether 'input' or 'response') and that to compare the trends one must sample from

---

Fig. 4.17   The application of spatial process-response analysis to the elevation of the Lower Greensand ridge in Southern England. (Chorley, 1969)

(a) The Lower Greensand outcrop and summit and ridge elevations.

(b) The best-fit quadratic surface fitted to the elevations shown in Fig. 4.17(a), explaining 93·52% of their variance.

(c) The best-fit quadratic surface fitted to the thicknesses of the Lower Greensand, explaining 89·44% of their variance.

(d) The best-fit quadratic surface fitted to the projected upper contact elevation, explaining 96·83% of the variance.

(e) The best-fit quadratic surface fitted to the percent silt and clay content of the surface soil, explaining 34·74% of the observed variance.

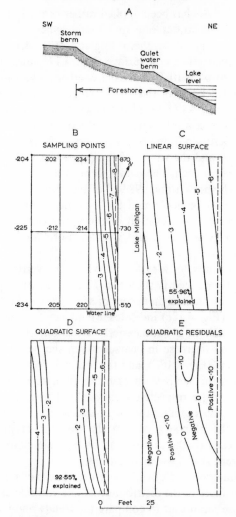

Fig. 4.18   The application of spatial process-response analysis to the median diameter of sand on a section of beach at Evanston, Illinois. (After Krumbein and Graybill, 1965)

(*a*) A generalized cross-section of the beach.

(*b*) The grid of sampling points, together with median grain sizes. Generalized grain-size isopleths are shown.

(*c*) The best-fit linear surface fitted to the median diameters, explaining 55·96% of the observed variance.

(*d*) The best-fit quadratic surface fitted to the median diameters, explaining 92·55% of the observed variance.

(*e*) The residuals from the best-fit quadratic surface.

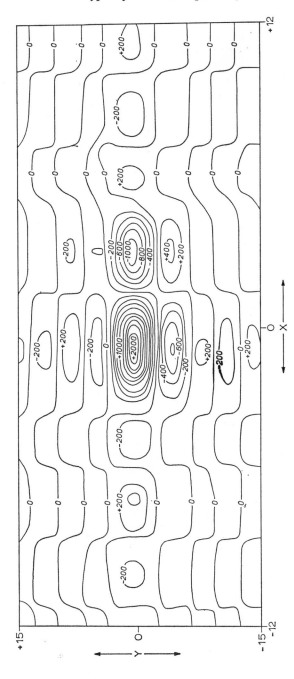

Fig. 4.19 A surface composed of synthetic double Fourier series containing four harmonics in each direction. (From Harbaugh and Preston, 1968)

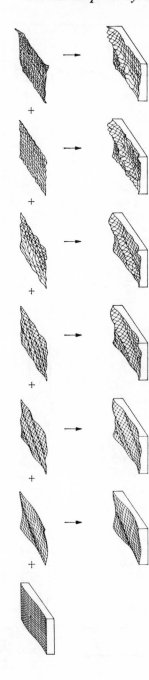

Fig. 4.20 Diagram illustrating the build-up of a complex terrain surface in Ghana by double Fourier synthesis. The final equation consists of some 321 real coefficients plus a residual. Only a few of the groups of terms are illustrated. (From Tobler, 1968)

these generalised surfaces and use multiple correlation techniques. This last step unfortunately introduces a great deal of error, for few trend surfaces have a 100% fit and, consequently, the point values obtained are approximations. There has been increasing realisation of the serious disadvantages inherent in the use of trend surfaces and as a result, attention is turning to a second method of space filtering.

The ideas behind space filtering have already been introduced in Chapter 2 and it should be clear that this method can be readily extended to the analysis of spatial processes, responses and their interrelationships. Central to the notion that rather complex spatial effects may be produced by the interaction of a number of simpler patterns which can be isolated and recognised by spatial filtering is this concept of decomposition of spatial responses. Trend surface analysis itself is one way of achieving this decomposition, but far more promising are techniques employing *harmonic analysis* (this is an extension of the analysis of *time series*, which is treated more fully in Chapter 7).

Briefly, just as it can be shown that any two-dimensional curve, however complex, can be successfully approximated by the superimposition of a number of simple, regular waves (commonly *Fourier* mathematical series containing sines and cosines), having various amplitudes, frequencies and phases; so the superimposition of *double Fourier series* can be used to build up complex surfaces. Fig. 4.19 shows a synthetic surface made up of double Fourier series containing four harmonics in each direction. The versatility of such surfaces in describing irregular terrain is illustrated in Fig. 4.20. In this way a complicated, three-dimensional response 'signal' is decomposed into its dominant harmonics and these, in suitable circumstances, can be used very effectively to elucidate the spatial operation of the processes responsible for the form of the surface. For example, a double Fourier analysis of an atmospheric pressure field might be expected to show the dominant airflow oscillations; and that of a landscape, such periodic features as fold belts and the hierarchy of drainage basin orders (each of which might be associated with a characteristic average basin width).

The great advantage of harmonic analysis over the use of trend surfaces is that, being derived from time series analysis, it is specifically designed for situations in which autocorrelation of sampling points is dominant. This means, of course, that one can have much greater confidence in the statistical reliability of the results of this technique. Both methods however, do suffer from the scale problem: at what level does one cease to 'bend' one's trend surface? or how many harmonics does one introduce into the Fourier analysis? In either case, one can progress to the point where the fitted surface exactly matches the real situation, with all its local variations. Generally, one halts the analysis at a coarser scale, but we must stress that the decision about

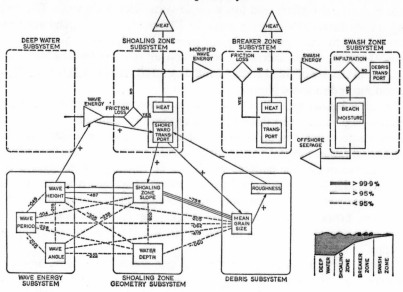

Fig. 4.21

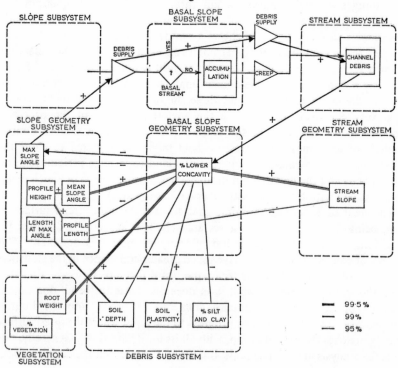

Fig. 4.22*a*

the cut-off point is always subjective and will depend upon the particular aims of each investigation.

4.5 Conclusion

In conclusion, we can see that all varieties of process-response systems are complex, but that there is a general increase in the problems of analysis as one progresses from point, to line systems, and through plane to irregular surfaces. Figs. 4.21–4.23 illustrate a number of attempts to indicate the nature of these complex structures that result from uniting morphological and cascading systems.

Fig. 4.21 takes certain of the beach zone structures previously described and forms them into a process-response system in which the cascade of wave

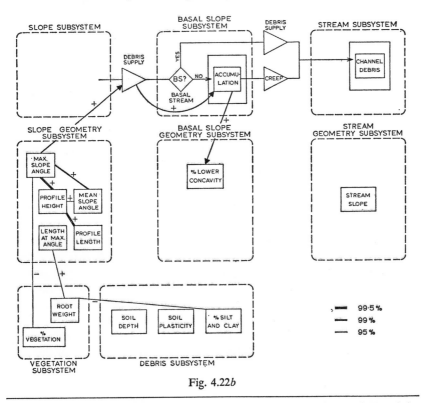

Fig. 4.22*b*

Fig. 4.21   A coastal process-response system. (Data from Harrison and Krumbein, 1964.) Values are those of '*r*'.

Fig. 4.22   A slope-stream process-response system on the Charmouthien Limestone, Plateau de Bassigny, Northern France. (Data from Kennedy, 1965).

(*a*) At locations where the basal stream is moving towards the valley-side slope.
(*b*) At locations where the basal stream is removed from the slope base.

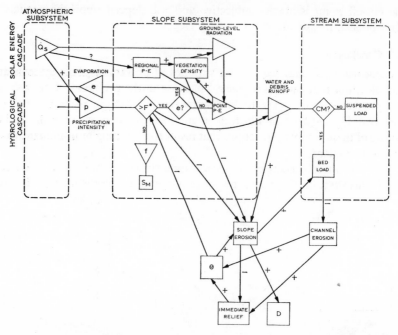

Fig. 4.23   A more complex process-response system involving two cascading systems and the morphometric properties derived from more than 50 drainage basins in the Western United States. (Data from Melton, 1957)

energy from deep water to the margin of the swash zone is linked to certain morphological features of the shoaling zone. (Of course this system could be made more complex and more realistic by building in the morphological properties of the breaker and swash zones in addition.) Fig. 4.22 performs a similar function for two valley-side slope and stream channel process-response systems, separated by a threshold in the local erosional environment, by linking the morphological properties of the slope to the debris cascade. Certainly the most complete process-response system investigated to date is that shown in Fig. 4.23, where the morphological variables of drainage basins in the western United States are linked to characteristics of the solar energy and hydrological cascades.

REFERENCES

Beckett, P. H. T. and Furley, P. A. (1968), Soil formation and slope development; *Zeitschrift für Geomorphologie*, **12**, 1–42.

Carson, M. A. (1969), Models of hillslope development under mass failure; *Geographical Analysis*, **1**, 76–100.

Chorley, R. J. (1969), The elevation of the Lower Greensand ridge, South-East England; *Geological Magazine,* **106,** 231–248.

Harbaugh, J. W. and Preston, F. W. (1968), Fourier series analysis in geology; In Berry, B. J. L. and Marble, D. F. Editors, *Spatial Analysis: A reader in statistical geography* (Prentice-Hall, New Jersey), 218–238.

Harrison, W. and Krumbein, W. C. (1964), Interactions of the beach-ocean-atmosphere system at Virginia Beach, Va.; *U.S. Army Coastal Engineer Research Center,* Technical Memo No. 7.

King, C. A. M. (1970), Feedback relationships in geomorphology; *Geografiska Annaler,* **52A,** 147–159.

Krumbein, W. C. (1964), A geological process-response model for analysis of beach phenomena; *Annual Report of the Beach Erosion Board for 1963,* **17,** 1–15.

Krumbein, W. C. and Graybill, F. A. (1965), *An Introduction to Statistical Models in Geology* (New York), 475.

Leclerg, E. L., Leonard, W. H. and Clark, A. G. (1962), *Field Plot Technique* (Burgess Publishing Co., Minneapolis), 373.

Melton, M. A. (1957), An analysis of the relations among elements of climate, surface properties and geomorphology; *Office of Naval Research Project NR 389–042, Technical Report 11* (Department of Geology, Columbia University, New York), 102.

Melton, M. A. (1958A), Geometric properties of mature drainage systems and their representation in an E_4 phase space; *Journal of Geology,* **66,** 35–54.

Melton, M. A. (1958B), Correlation structure of morphometric properties of drainage systems and their controlling agents; *Journal of Geology,* **66,** 442–460.

Ploey, J. de and Savat, J. (1968), Contribution à l'étude de l'érosion par le splash; *Zeitschrift für Geomorphologie,* **12,** 174–193.

Tobler, W. (1968), *Geographical Filters and their Inverses* (Mimeo), (Department of Geography, University of Michigan), 16, plus appendix.

Towler, J. E. (1969), *A Comparative Analysis of Scree Systems Developed on the Skiddaw Slates and Borrowdale Volcanic Series of the English Lake District;* (B. A. Dissertation, Department of Geography, Cambridge University), 84.

Whitten, E. H. T. (1964), Process-response models in geology; *Bulletin of the Geological Society of America,* **75,** 455–464.

5: *Input and Output*

A system is a big black box
Of which we can't unlock the locks,
And all we can find out about
Is what goes in and what comes out.

Perceiving input-output pairs,
Related by parameters,
Permits us, sometimes, to relate
An input, output, and a state.
If this relation's good and stable
Then to predict we may be able,
But if this fails us—heaven forbid!
We'll be compelled to force the lid!

KENNETH BOULDING

5.1 Introduction

Many natural events constitute inputs, throughputs and outputs of energy and mass to and from cascading systems and subsystems. As inputs these events may be largely independent of the system organization, but the latter may be temporarily or permanently modified by the input if it exceeds a given threshold. As system outputs or throughputs between subsystems, however, the character (i.e. the magnitude and frequency) of these events may be partly conditioned by the structure of the cascading systems through which they have passed. Thus the magnitude and frequency of solar energy incident upon a given area; the amount, duration and intensity of precipitation falling on a given drainage basin; and the height and frequency of waves approaching a beach; are all determined by processes external to the area, basin or beach in question. As soon as the energy and mass enters the system, however, its disposition is partly determined by properties inherent in the system, and the receipt or reflection of solar energy, the infiltration or surface runoff of rain in a drainage basin and the dissipation or transfer of wave energy in the zone of shoaling waves all depend partly on the morphological qualities of the systems themselves. This action is subject to feedback and, for example, excessive solar energy can change the surface characteristics by causing vegetation amounts and types to change, intense rainstorms can erode the surface and so alter the relation of infiltration to runoff, and large waves can scour the bottom of the zone of shoaling waves and permit a subsequently greater transfer of wave energy shoreward. The effect of this complicated interrelationship of input to throughput is to transform patterns of input

160

events into different patterns of outputs, which may themselves cascade as inputs into other subsystems. Any treatment of input and output therefore involves consideration of the inherent characteristics of inputs, the modifications imposed on these by the system characteristics producing differing output characteristics, and the capacity of given energy distributions to overcome the strength thresholds of the system morphology, thereby transforming the system itself.

5.2 Probability Distributions of Events

The occurrence in time of energy events has long been subject to observation, and Fig. 5.1 shows as a bar graph (*histogram*) the distribution of maximum daily temperatures recorded in Washington, D.C. for the 6808 days of June, July and August for the period 1872–1945. It is clear that the large number of energy events presents a distinctly symmetrical spread around the most common maximum summer day temperatures in the mid 80°'s F. This type of sample distribution is encountered so commonly in nature that it is termed a *normal distribution*. The assumption underlying the association of a histogram made up of a sample (of N events or *variates*, X_i) is that if measurement could continue until an infinite number of events were included, and the class intervals in which they are grouped were made infinitely small, the percentage histogram would become a smooth, bell-shaped curve. Of course, the assumption that the total population of given events (e.g. maximum daily summer temperatures in Washington, D.C.) is infinite is questionable, as we shall see later, but the association of many samples with appropriate normal distributions has proved a powerful predictive tool in investigating many natural phenomena. The characteristics of the normal curve appropriate to a given frequency histogram of events can be expressed by means of a number of *parameters*, or diagnostic numerical values.

The value of X about which all other values are assumed to be symmetrical (e.g. 84·5°F in Fig. 5.1) can be expressed by three *measures of central tendency:*

1 The *arithmetic mean* ($\bar{X}$); or average value around which the others balance, where $\bar{X} = \left(\sum\limits_{i=1}^{N} X_i \right) / N$

2 The median ($\tilde{X}$); or middle value when the variates are ranked in order of magnitude. Areas under the frequency distribution curve are equal on either side of the median. *Quartiles* (Q_{25} and Q_{75}) further subdivide the distribution into four parts each with an equal number of variates.

3 The mode ($\hat{X}$); or most frequent value is that above which the peak in the frequency distribution occurs.

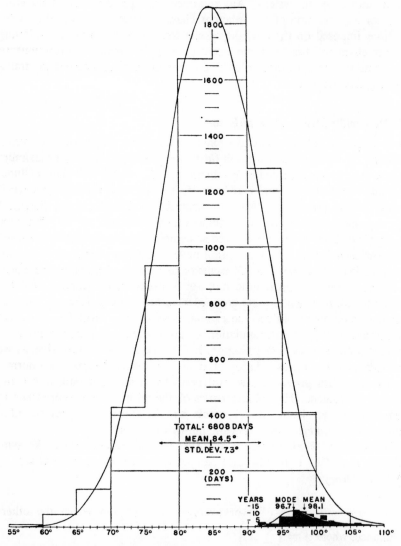

Fig. 5.1   Frequency of highest temperatures for each summer day (and for each year in black) for Washington, D.C. during the months June to August for the 74-year period 1872–1945. (From Court, 1952)

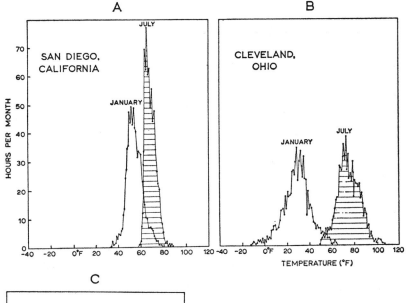

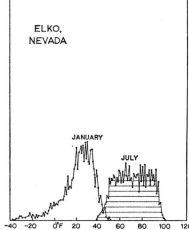

Fig. 5.2   January and July hourly temperature frequencies for (*a*) San Diego, (*b*) Cleveland and (*c*) Elko during the period 1935–1939 (i.e. approximately 3700 observations for each station). (After Court, 1951)

For the normal distribution it is clear that $\bar{X} = \tilde{X} = \hat{X}$, but this is not so for other distributions. (Nor is a distribution with $\bar{X} = \tilde{X} = \hat{X}$ necessarily normal.)

Another characteristic of the distribution of events is the amount of spread or *dispersion* of the variates about the mean. For normal distribution this is expressed in terms of the standard deviation (σ), where

$$\sigma = \sqrt{\sum_{i=1}^{N} [(X_i - \bar{X})^2]/N}$$

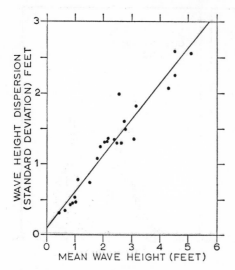

Fig. 5.3   The relation of standard deviation of wave heights to mean wave height. (After Putz, 1952)

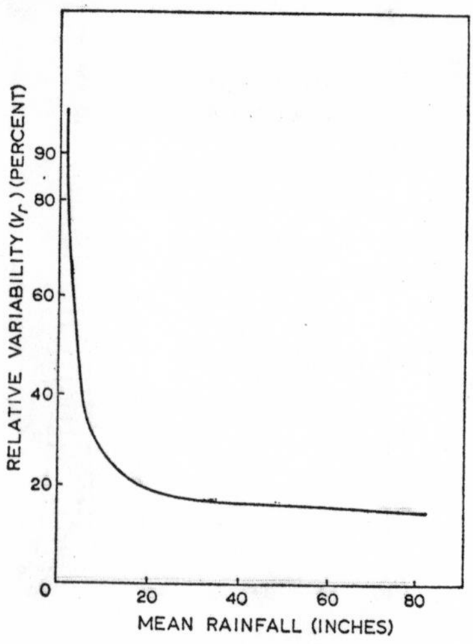

Fig. 5.4   The generalized relationship between relative variability of rainfall (V_r) and mean rainfall. (After Conrad and Pollack, 1950)

The standard deviation has the same units as that of the events and has the property that, for a normal distribution, one can predict that 68·27% of all the events will lie between ±1σ around the arithmetic mean, 95·45% between ±2σ, and 99·73% between ±3σ. For a normal distribution, 50% of the events lie between plus or minus one *probable deviation* (=0·6745σ) around the mean. In Fig. 5.1, for example, the value of σ is 7·3°F. Thus normal distributions with identical mean values can have different standard deviations. Frequency distributions of hourly temperatures in individual months for given stations appear to be normally distributed with winter standard deviations greater than those in summer months and showing a general increase of σ as one passes from maritime locations (Fig. 5.2a), to humid inland stations (Fig. 5.2b), to arid inland stations (Fig. 5.2c). It is a general characteristic of natural events that the greater the mean the greater the standard deviation (Fig. 5.3). Dimensionless measures of the dispersion of events have been devised, including:

$$\text{Relative variability } (V_r) = \left(\sum_{i=1}^{N} (X_i - \bar{X})/N \right)$$

$$\text{Coefficient of variation } (V_c) = [\sigma/\bar{X}] \,.\, 100$$

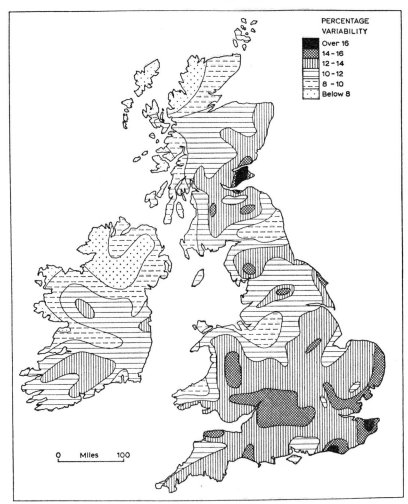

Fig. 5.5   The relative variability of annual rainfall over the British Isles (1901–1930).
(After Gregory, 1963)

It is a common observation that relative variability (V_r) decreases as the
mean increases; Fig. 5.4 illustrates this for mean annual rainfall, and Fig. 5.5
maps the relative variability of annual rainfall over the British Isles.

The normal distribution (Fig. 5.6a) can be transformed into a much more
usable linear form by first plotting it in a cumulative form (Fig. 5.6b) and
then transferring it to *probability paper* (Fig. 5.6c) on which the vertical scale
has been distorted so that cumulative normal distributions appear as straight

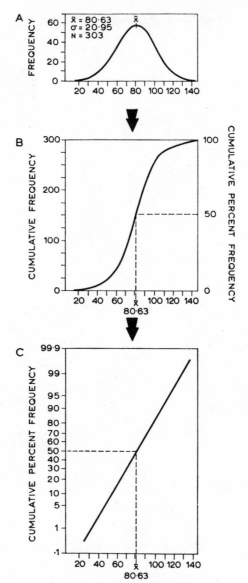

Fig. 5.6   Stages in the transformation of a normal distribution (*a*) successively into a cumulative form (*b*) and then into an arithmetic probability (*c*).

Fig. 5.7   The use of arithmetic probability plots to estimate the normality of distributions relating to:

(*a*) Annual precipitation for selected U.S. stations. (From Hershfield, 1962)

(*b*) Annual discharge of the Colorado River at Lees Ferry (1896–1956). The probable deviation is equal to 0·6745 times the standard deviation. In this instance the latter is equal to 4·22 million acre-feet. (From *Fluvial Processes in Geomorphology* by Luna B. Leopold, Gordon Wolman and John P. Miller. W. H. Freeman & Company. Copyright © 1964)

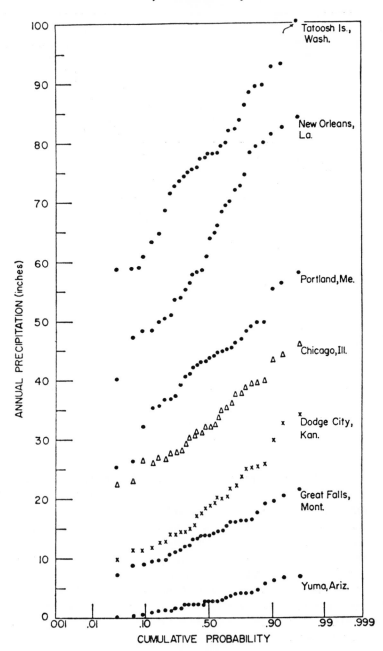

Fig. 5.7*a*

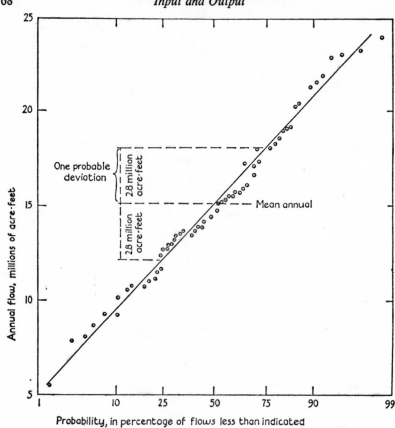

Fig. 5.7*b*

lines, each line having a slope which is directly proportional to the standard deviation. Fig. 5.7*a–b* shows how the normality of samples of annual precipitation and annual river discharge events can be estimated by employing such probability plots, although, of course, more rigorous tests for normality are in use. A special feature of probability plots is that one can deduce the percentage of events which are less than a given magnitude, and therefore the probability of encountering an event of that magnitude. This is especially important in studies of inputs into systems. Using normal curves fitted to annual station rainfall data, for example, it is possible to estimate the probability (*p*) that a variate in a normal distribution will exceed a given value, by calculating the statistic '*u*' where:

$$u = (X_i - \bar{X})/\sigma$$

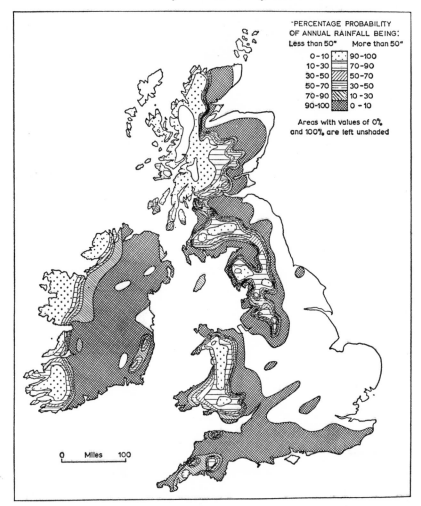

Fig. 5.8   The percentage probability of receiving an annual rainfall of less, or more, than 50 inches in the British Isles. (After Gregory, 1957)

and then by using u to obtain p in a table of areas under the normal curve. Fig. 5.8 maps for the British Isles the percentage probability of receiving an annual fall of less, or more, than 50 inches.

Unfortunately, most distributions of process events constituting inputs, throughputs and outputs in physical geography are *right-skewed*, leading to differences between the mode, median and mean (Fig. 5.9) such that the

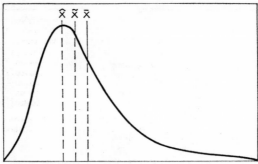

Fig. 5.9   The difference between the arithmetic mean ($\bar{X}$), the median ($\tilde{X}$) and the mode ($\hat{X}$) of a right-skewed distribution.

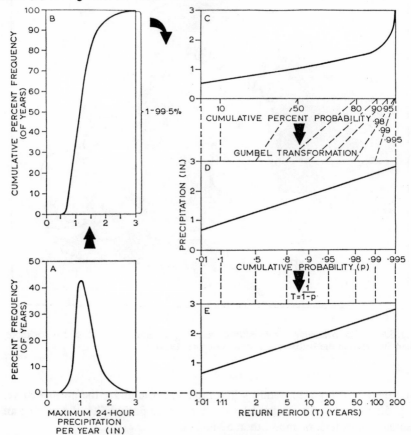

Fig. 5.10   Stages in the transformation of a right-skewed distribution (*a*), into a cumulative form (*b*), rotated and slightly abbreviated (*c*), subject to a Gumbel transformation (*d*), and expressed in terms of the return period (*e*).

degree of skewness is given by $(\bar{X} - \hat{X})/\sigma$. These skewed distributions present certain problems of description and of inferring probabilities for them. When plotted on arithmetic probability paper right-skewed distributions appear as convex curves (Fig. 5.10*a* and *b*: we shall return to the discussion of Fig. 5.10*c*, *d* and *e* later.) A moderate amount of right-skewness can be normalized by making a logarithmic transformation of each of the variates or by plotting the cumulative frequency on *logarithmic probability paper*. A variety of energy events in nature have been approximated by these *logarithmic normal* distributions, including rates of soil creep (Fig. 5.11*a*) and the occurrence of daily wave heights (Fig. 5.11*b*). However, the range of skewness, even of similar events, is large, particularly in the cases of rainfall (Fig. 5.12) and stream discharge, and more general treatments of the whole family of right-skewed events have proved necessary. There are a number of influences which promote this characteristic right-skewness of recorded natural events:

1 Where the magnitude of given events is absolutely limited at the lower end (i.e. it is not possible to have less than zero rainfall or runoff), or is effectively so (i.e. as with low temperature conditions), and not at the upper end, the infrequent events of high magnitude cause the characteristic right-skew.

2 The above-mentioned limitation of the lower magnitudes implies that as the mean of the distribution approaches this lower limit, the distributions become more skewed (Fig. 5.12).

3 The longer the period of record, the greater the probability of observing infrequent events of high magnitude, and consequently the greater the skewness.

4 The shorter the time intervals within which measurements are made, the greater the probability of recording infrequent events of high magnitude and the greater the skewness. Fig. 5.1 shows, for example, how much more skewed is the distribution of the 74 annual highest daily maximum temperatures at Washington, D.C. than is the distribution of the 6808 maximum daily summer temperatures for the same 74 year period. Fig. 5.13*a* demonstrates the same principle by comparing the distributions of annual precipitation (almost normal) and January precipitation (strongly right-skewed) for Hilo, Hawaii.

5 Other physical principles tend to produce skewed frequency distributions of events. For example, the limited size of high intensity thunderstorms means that the smaller the drainage basin the higher the probability that it will be completely blanketed by heavy rain, and this leads to an increase

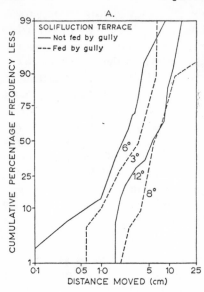

Fig. 5.11   Cumulative percentage frequency distributions of events plotted on logarithmic probability graphs.

(*a*) Mass movements on solifluction terraces on Ben Lomond, Tasmania (November 1963–February 1964), indicating the skewed nature of the frequency of events, and that maximum rates of movement are observed on steeper slopes fed by headward gully systems. (After Caine, 1968)
(*b*) Maximum daily wave heights at Cleveland (Lake Erie).

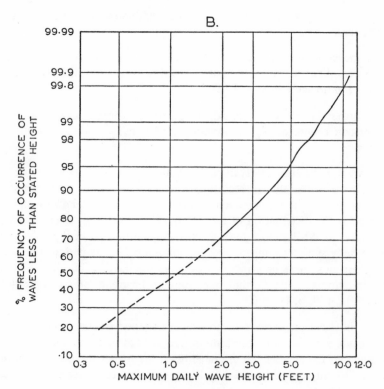

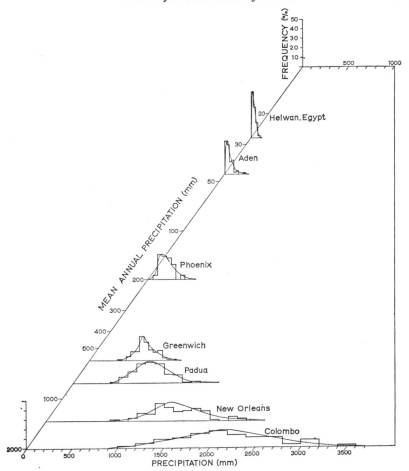

Fig. 5.12   Frequency distribution curves of annual precipitation amounts for seven stations.

of skewness in the distribution of runoff as basin size decreases. Similarly, stream discharge frequencies are extremely skewed where impermeable strata allow little ground-water runoff and the frequency of low flows is restricted.

The problems presented by the skewness of natural events are made more complex by the variability of the amount of skew. On a simple level, distributions of mean annual temperatures, rainfall and stream discharge give mean values of skewness of 0·04, 0·09 and 0·23, respectively. However, even given single events exhibit differing degrees of skewness depending on

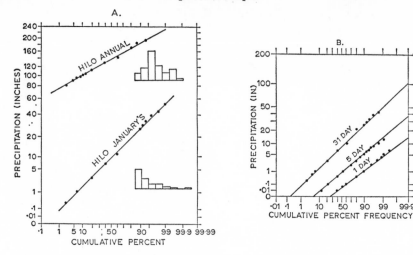

Fig. 5.13   Frequency distributions of precipitation on Hilo, Hawaii. (After Stidd, 1953)
(*a*) Annual and January precipitation data plotted on a cube-root probability graph.
(*b*) 1-day, 5-day and 31-day precipitation data for January plotted on a cube-root probability graph.

the temporal or spatial bases on which they are observed. Attempts to rationalize such distributions have been made in various ways in order to make some prediction of their frequency of occurrence. The logarithmic transformation has already been mentioned, and the cube root transformation has proved successful in normalizing rainfall data with the slope of the cumulative frequency plot directly related to the skew. Fig. 5.13*a* shows such plots on cube-root probability paper and their applicability to the description of short period rainfall distributions is shown in Fig. 5.13*b*.

There are, however, two drawbacks to the normalization methods previously described, firstly, they appear to be effective over only limited ranges of skewness, and, secondly, their use is not supported by any abstract theoretical basis. Basing his work on the *statistics of extreme values*, E. J. Gumbel showed that such right-skewed distributions of natural events can be considered as extreme values of large populations of independent events. Employing this theory, wide ranges of skewness can be characterized as a family and a special *extremal probability paper* has been designed to more-or-less normalize many natural distributions (Fig. 5.10*c–e*). The main difference between this co-ordinate system and those of the logarithmic and cube-root normal distributions is that the scale on which the variate values are plotted (vertical) is unaltered, but the probability scale (horizontal) has been transformed in accordance with the theory of extremes. It should be noted,

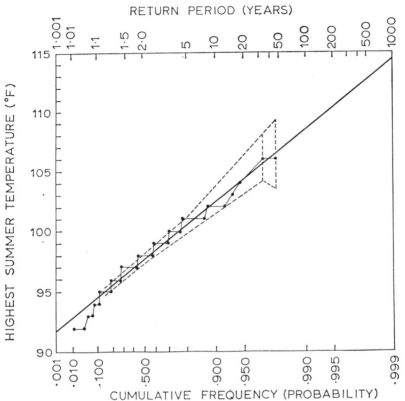

Fig. 5.14   Extremal probability plot of the highest summer temperatures at Washington, D.C. (1871–1945) shown in the small histogram of Fig. 5.1. The 95% confidence band is included. (After Court, 1952)

however, that where very extremely skewed distributions are encountered, a logarithmic scale for the variates is sometimes used. The use of extremal probability paper to straighten out cumulative plots of highly skewed distributions is shown for the highest summer temperatures at Washington, D.C. shown in Fig. 5.1 (Fig. 5.14) and for maximum daily rainfalls at Baltimore (Fig. 5.15). In practice this method appears to work quite well although a number of criticisms have been levelled at it:

1 Natural events such as temperature, rainfall and runoff, no matter what time scale they are measured on, are not truly independent events. A later section will show a clustering tendency, opposed to randomness, for such events.

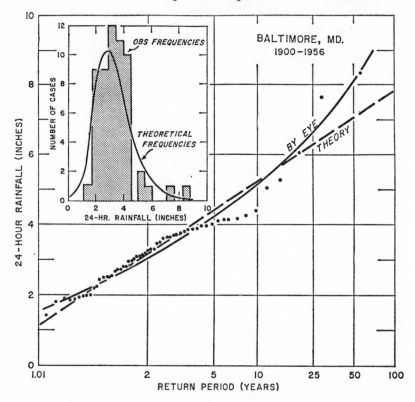

Fig. 5.15   Extremal probability plot of annual maximum 24-hour precipitation data for Baltimore, Maryland (1900–1956). (From Hershfield and Kohler, 1960)

2 The number of such extreme events may not be large, as demanded of theory, especially if—

3 Climatic changes are continually changing the character of the population from which extreme values are being drawn.

5.3 Probability and Return Periods

The basic aspects of the statistics of extremes can be expressed in a very simple manner. If an event of given magnitude (X) is observed to occur H times in N time intervals (i.e. a discharge which just overtops a river bank 5 times in 25 years), then the relative frequency of occurrence is H/N (i.e. 5/25) and the mean *return period* (T) is N/H (i.e. 25/5 = 5 years). This average return period does not imply, of course, that an event of magnitude X will occur regularly every 5 years, but for a large number of these events the average period of time separating them is 5 years. In fact very often the river bank may be overtopped in successive years. If the probability of the occurrence of an event of *at least* as great a magnitude as X in one time interval is p, then its probability of non-occurrence or 'exceedence' (q) in the same interval is $1-p$. Now, the mean return period can be expressed simply in terms of the probability of occurrence (Fig. 5.10 *d* and *e*):

$T = 1/p$ (in this case $= 1/0\cdot2 = 5$ years)

$\therefore T = 1/(1-q)$

$\therefore q = 1 - (1/T)$

The probability of the occurrence of an event of at least the magnitude of X in N time intervals is given by the expression

$1-q^N = 1-[1-(1/T)]^N$

If N is expressed as a fraction $(1/r)$ of T such that

$$N = T/r$$

then substituting above:

$$1-q^N = 1 - [(T - 1)/T]^{T/r}$$

As T approaches infinity for a large sample or a long time, the right hand side of this equation approaches $1 - e^{-1/r}$, where e is the base of natural logarithms (2·71828). From this relationship many key parameters relating to T can be deduced, such that the mean value of T ($\bar{T}$) is 2·33 time intervals and the most probable value of T (the mode $= \hat{T}$) is 1·58 time intervals (i.e. it is more likely that an event of given order of magnitude is immediately followed by a similar event than by one of a very different order of magnitude). If, for example, an annual event has a probability ($p = 0.01$) of occurring once in a hundred years (i.e. $T = 100$ years), the theory of the statistics of extremes indicates that of the 10,000 times it is to be expected in 1 million years, 50% of the time intervals separating successive events will be less than 70 years and 1% of the events will be expected to occur on successive years. Putting things another way, for an event with an average return period of 100 years, there is a probability of 0·63 that it will occur or be exceeded in any given 100-year period, 0·50 in any 70-year period and 0·40 in any 50-year period. (The rationale of this is that if an event has an average return period of T, the probability is $e^{-N/T}$ that it will not occur or be exceeded in less than N years.) Before leaving this brief excursion into the theory of the statistics of extremes, it should be noted that, partly for the reasons given above, actual events depart in detail from theoretical predictions. In particular, events of large

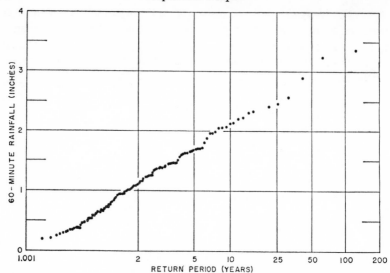

Fig. 5.16　　Extremal probability plot of annual maximum 60-minute precipitation data for 128 United States' stations in 1948. (From Hershfield and Kohler, 1960)

return periods show especially random tendencies, and it is a common feature of extremal probability plots for actual records (e.g. of flood discharges) that they give magnitudes smaller than those to be inferred from straight-line probability plots.

In practice it is possible to obtain results essentially identical to those produced by the application of the theory of extreme statistics by adopting the following simple ranking procedure:

1　Arrange the N values of magnitude in descending order (M, where $M = 1$ is the largest and $M = N$ is the smallest).
2　Calculate T, from the formula $T = (N + 1)/M$ (e.g. in years).
3　Plot T against magnitude on extremal probability paper.
4　Fit a straight line to the plot by eye.

Extremal probability plots have been constructed for many natural events including rainfall (Figs. 1.6 and 5.16), annual temperature maxima (Fig. 5.17), the peak river discharges recorded each year (Fig. 5.18a) and mean annual discharges (Fig. 5.18b). The dashed lines in Fig. 5.18 indicate confidence limits, which are calculated in the same way as for an ordinary regression line.

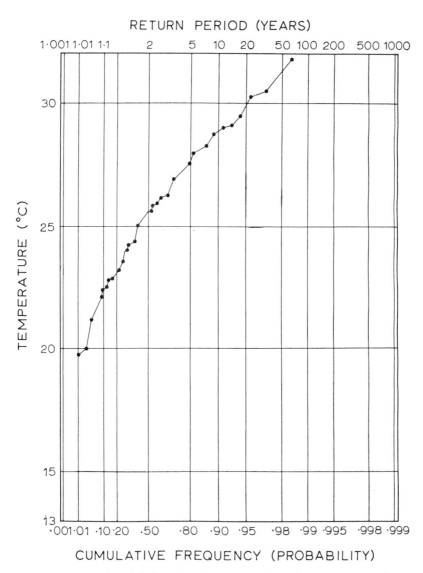

Fig. 5.17   Extremal probability plot of annual maxima of temperatures at Bergen, Norway (1857–1926). (After Gumbel, 1958A)

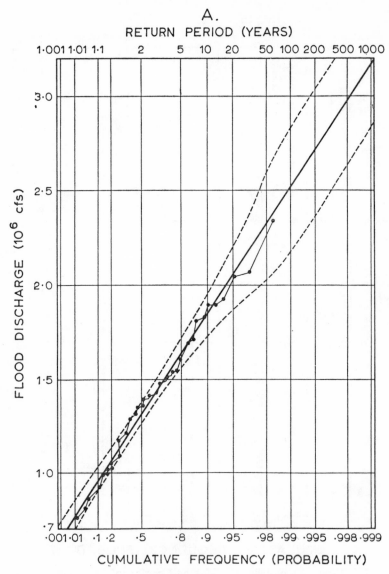

A.

Fig. 5.18a

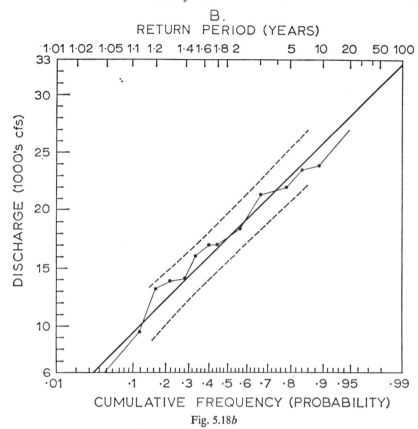

Fig. 5.18*b*

Maps of events of given magnitude and frequency can be prepared for temperature (Fig. 5.19), precipitation (Fig. 5.20 and 5.21) and floods (Fig. 5.22).

Of course, it has always proved tempting to use extremal plots as a means of extending inference beyond the existing record and to attempt the estimation of the magnitude of events whose frequency is greater than the length of record. Such inferences are extremely useful for engineering planning, but the statistical difficulties encountered at the upper end of extremal plots make such extrapolation hazardous.

Fig. 5.18   Extremal probability plots of stream discharges (After Gumbel, 1958A). 95% confidence bands are shown.

(*a*) Annual peak flood discharges for the Mississippi River at Vicksburg, Mississippi (1898–1949).
(*b*) Annual mean discharges for the Colorado River at Bright Angel Creek (1922–1939).

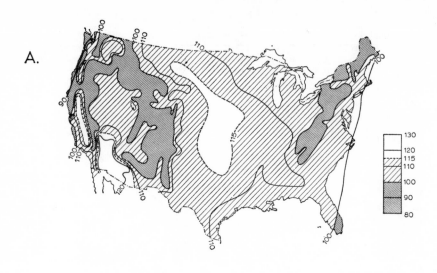

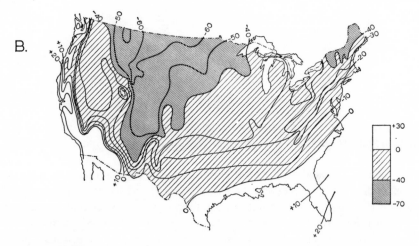

Fig. 5.19   Highest (*a*) and lowest (*b*) temperatures (°F) to be expected to occur once in 100 years in the United States. (After Court, 1953.) Based an analyses of tempearture data for the period 1901–1930.

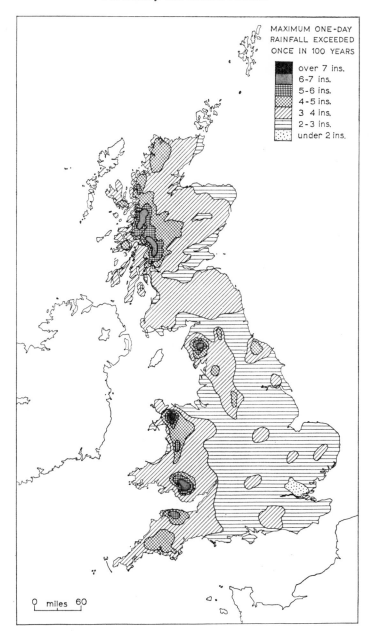

Fig. 5.20   Maximum one-day rainfall which may be expected to be exceeded once in 100 years for Great Britain. (After Rodda, 1967)

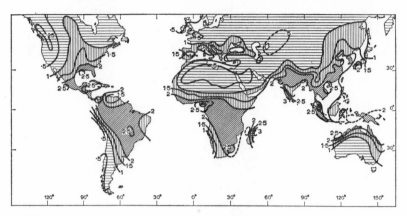

Fig. 5.21   Tentatively-proposed world map of the 2-year/1-hour maximum precipitation. (After Reich, 1963)

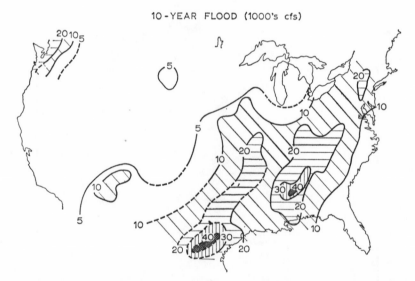

Fig. 5.22   The 10-year flood potential for the United States, in 1000s cfs, from drainage basins of 300-square-mile area. (From *Fluvial Processes in Geomorphology* by Luna B. Leopold, Gordon Wolman and John P. Miller. W. H. Freeman & Company. Copyright © 1964)

5.4 Inputs and Throughputs

So far we have assumed that energy events are random and causally unconnected, but of course this is not so. Natural events fall between two extreme models:

1 The deterministic (or 'long memory') model, where a given event can be completely predicted on the basis of the events leading up to it.
2 The stochastic (or 'no memory') model, where events appear to be quite random.

Natural events are of an intermediate, or 'short memory', type and are best exemplified by the *Markov chain*. The Markov chain is a sequence of states or events in which each feature of any given event (for example, its magnitude) is related partly to the condition of a previous event and is partly random. The *order of the chain* is dependent upon the proximity of the previous related event, and, for example, a first-order Markov chain refers to related events immediately preceding in time those with which one is concerned. For our purposes, therefore, a Markov chain is a sequence of events (1, 2, 3, ... n) which is associated with a matrix of transition probabilities (p_{ij}) giving information about the behaviour of the system in terms of the probability of an event 'i' being succeeded at one or more remove by a state 'j', thus:

			SUCCEEDING EVENT				
				(j)			ROW
			1	2	3	4	TOTAL
PREVIOUS EVENT		1	P_{11}	P_{12}	P_{13}	P_{14}	1·00
	(i)	2	P_{21}	P_{22}	P_{23}	P_{24}	1·00
		3	P_{31}	P_{32}	P_{33}	P_{34}	1·00
		4	P_{41}	P_{42}	P_{43}	P_{44}	1·00
	COLUMN TOTAL		1·00	1·00	1·00	1·00	

So, for example, P_{34} is the probability of an event of magnitude 3 being followed by one of magnitude 4. All rows and columns of the transition probability matrix total 1·00—i.e. there is absolute certainty, for example, that an event of magnitude 1 will be followed by an event of magnitude 1, 2, 3, or 4.

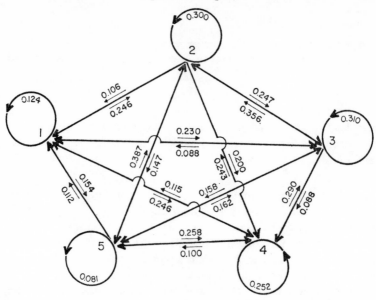

Fig. 5.23   Kinematic graph of transition probabilities for annual river discharges in the Potomac Basin (1930–1960). The discharge classes are described in the text. (From Chorafas, 1965)

A simple example of a Markov analysis of river discharge events has been conducted from the 553 total years of record of 29 rivers in the Potomac basin. These annual discharges were divided into 5 classes (Table 5.1) and the occurrence and transition probability matrices constructed. Now, if annual

TABLE 5.1

Discharge Classes	Per cent of Mean Annual Discharge
Class 1 (much below average)	less than 60%
Class 2 (below average)	60–89%
Class 3 (average)	90–109%
Class 4 (above average)	110–139%
Class 5 (much above average)	140% and up

discharge events were random and unconnected one might expect that a sufficiently large number of events divided into 5 classes would give a random probability matrix with all values equal to 0·200 (i.e. 20%). In the present case this is not so, and, as the *kinematic graph* of transition probabilities

(Fig. 5.23) shows, for example, that a year of average flow has a 66·6% probability of being followed by annual flows of 'average' or 'below average' magnitude—as against a 40% probability if the sequences were random. An interesting extension of this Markovian analysis is to use the result to generate synthetic sequences of events which can be compared with actual sequences (Fig. 5.24).

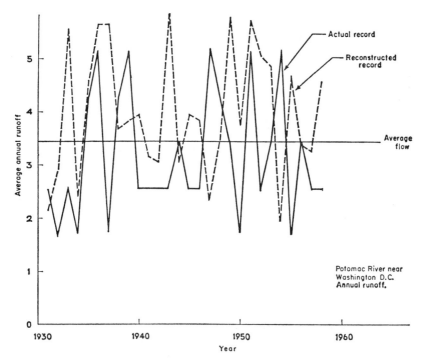

Fig. 5.24   Graphs of actual and simulated sequences of annual river discharges in the Potomac Basin (1930–1960). (From Chorafas, 1965)

Of course, it is obvious that the mechanisms of such energy events as rainstorms, daily stream discharges, individual wave and daily temperature magnitudes imply probability relationships between individual events and those following shortly afterwards. The control by depression systems and other influences which encourage sequences of weather events is shown in Fig. 5.25 where curves of probability (p) that a given day of the year (t) will be dry ($P(D_t)$) are compared with first- and second-order Markov plots, where $P(D_t/D_{t-1})$ is the probability that day t will be dry, if day $t-1$ was

14

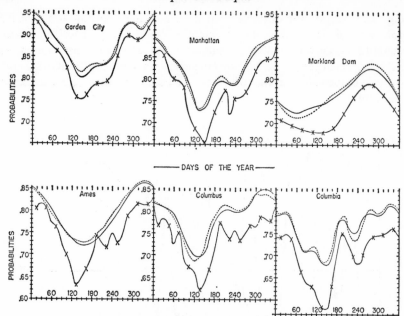

Fig. 5.25 Probabilities of dry-day (i.e. <0·01 inch precipitation) sequences for six stations in the United States. (From Feyerherm and Bark, 1965)

—X—X— Probability that a given day of the year will be dry.
———— Probability that a given day will be dry if the preceding day was also dry.
— — — — Probability that a given day will be dry if the two preceding days were also dry.

also dry. Table 5.2 gives a third-degree Markovian transition probability matrix relating to diurnal rainfalls for 27 rainy seasons (November–April 1923–1950) in Tel Aviv, Israel according to the particular three-day sequence that preceded a given wet day ($W \geqslant 0·1$ mm rainfall).

TABLE 5.2

i-2	i-1	i	W	j D	Row Total
W	W	W	0·654	0·346	1·000
D	W	W	0·610	0·390	1·000
W	D	W	0·674	0·326	1·000
D	D	W	0·720	0·280	1·000
W	W	D	0·227	0·773	1·000
D	W	D	0·307	0·693	1·000
W	D	D	0·266	0·734	1·000
D	D	D	0·245	0·755	1·000

(Gabriel and Newmann, 1962.)

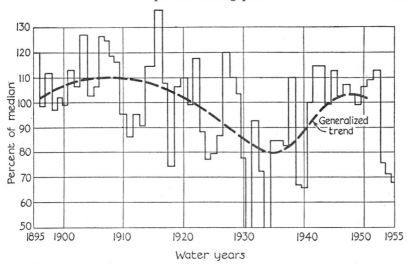

Fig. 5.26   Variation in annual runoff generalized for the United States as a whole (1895–1955). The dashed line suggests a cyclic trend. (From *Fluvial Processes in Geomorphology* by Luna B. Leopold, Gordon Wolman and John P. Miller. W. H. Freeman & Company. Copyright © 1964)

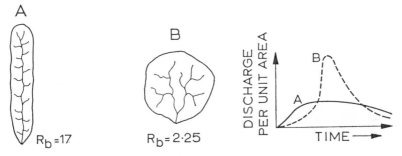

Fig. 5.27   The influence of basin shape on stream runoff hydrographs, in respect of basins having (*a*) high and (*b*) low bifurcation ratios. (After Strahler. From Haggett and Chorley, 1969)

Such work, together with more complicated analyses of weather sequences has demonstrated that in most areas of the world energy events occur in 'spells' such that a high intensity event has a better chance of being succeeded by an above-average than a below-average event. In hydrology the concept of the average recurrence interval is complicated by this *persistence effect*, with events of high and low magnitude showing some evidence of segregated grouping. This has been shown for almost 1000 years of Nile discharge figures. One method of identifying the grouping of such spells of events is

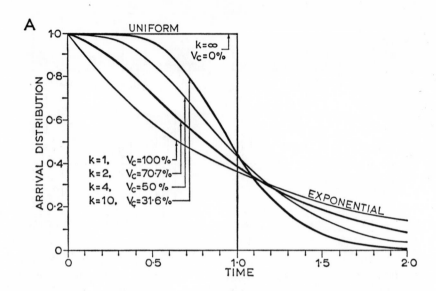

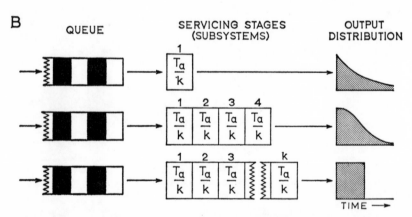

Fig. 5.28   Queue distributions (After Gordon, 1969). A full description is given in the text.

(*a*) Erlang distributions of certain types of telephone traffic.
(*b*) Output distributions resulting from the servicing of continuous queues of arrivals by k stages (subsystems), each having an exponentially-distributed service time.

provided by the recognition that the arithmetic means of groups of non-random events show a greater range (variance) than those of random events. Indeed, all temporal scales of such sequences—daily, weekly, seasonal, year-to-year, etc.—have been identified, and separating clustered sequences of no general trend from cyclical or periodic variations (Fig. 5.26), and cyclic variations from large-scale significant changes are difficult operations (see Chapter 7).

Even if one has knowledge of the magnitude and frequency of system inputs, the complex nature of process-response systems implies the transformation of the character of an energy pulse as it moves through the system. This is exemplified by the relationship of the output stream hydrograph to the input storm rainfall (see Fig. 1.7), and by the effect of basin geometry on discharge outputs resulting from uniform rainfall inputs (Fig. 5.27). When inputs are closely spaced in time their transmission through the system is often blurred and confused, particularly as the basin hydrological cycle represents a complex queueing problem where orderly 'queues' of water are distributed by filled storages and by-passing. Attempts have been made, for example, to treat soil moisture disposition as a queueing problem where assumed inputs of rainfall of given time duration and magnitude distribution are queued into soil storages and 'serviced' by evapotranspiration and other moisture losses. The arrival pattern of items in a queue can be described in terms of the *inter-arrival time* (i.e. the time interval between successive arrivals, where T_a is the mean inter-arrival time) such that *arrival distributions* can be defined in terms of the probability that an inter-arrival time is greater than a given time. A. K. Erlang has shown that certain types of telephone traffic can be represented by a family of curves (Fig. 5.28a) in which the value of the constant k is inversely related to the coefficient of variation (V_c) (see Chapter 5.2). Where $k = 1$ the distribution is exponential with $V_c = 100\%$, and when $k = \infty$ ($V_c = 0\%$) the distribution is uniform and the inter-arrival time constant. It is possible to use these *Erlang distributions* to rationalize output distributions from a series of servicing stages (subsystems) and Anderson has suggested an application of this idea in relating basin runoff to rainfall distributions. If a continuous queue of arrivals presents itself for servicing through k stages or subsystems, each having an exponentially-distributed service time with the same mean value (T_a/k) (Fig. 5.28b) such that each item is subjected to a random and independently-selected service time at each stage with no arrival being allowed to enter the first stage until its predecessor has cleared the last stage, the distribution of over-all service times (i.e. the output distribution) is an Erlang distribution of the kth order with a mean of T_a. Thus Fig. 5.28 can be viewed as a throughput model in which the value of k of the output distribution gives an indication of the number of subsystems which have serviced the throughput.

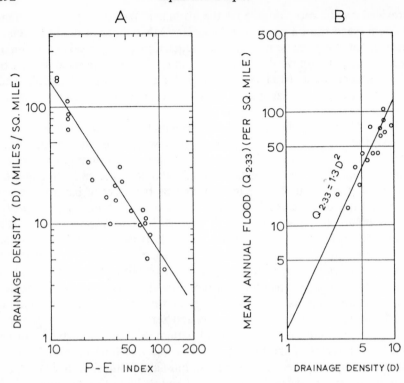

Fig. 5.29   The relationships between drainage density and two hydrological variables. (From Chorley, 1969)

(*a*) The control over drainage density exercised by Thornthwaite's precipitation-effectiveness (P–E) index. (From Melton, 1957)

(*b*) The control over mean annual flood ($Q_{2.33}$) exercised by drainage density for 13 basins in the central and eastern United States. (After Carlston)

The basic problem of the study of system throughputs involves the fundamental question: Under what conditions does a given input passively become a throughput and output, and under what conditions it is of sufficient magnitude to perform significant work to transform the system itself? Chapter 7 will return to this question in relation to the importance of time-scales, and it will suffice now to point out that it is the different basic reference time scales of the geomorphologist and hydrologist which lead them to view runoff either as a cause or an effect of drainage density (Fig. 5.29). It is necessary here, however, to consider the nature of the work done on systems by their inputs.

5.5 Work

Obviously all energy throughputs have some effect in producing change, degradation or reorganization in systems, although these effects vary greatly with the magnitude of the event and are often regulated by the existence of

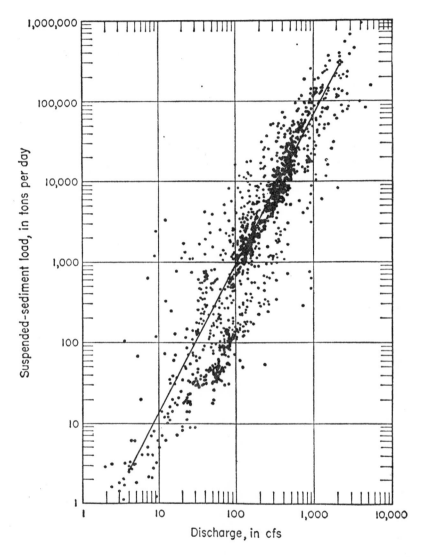

Fig. 5.30   The logarithmic relationship of suspended sediment load to discharge for the Powder River at Arvada, Wyoming. (From Leopold and Maddock, 1953)

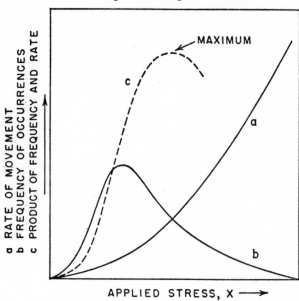

Fig. 5.31

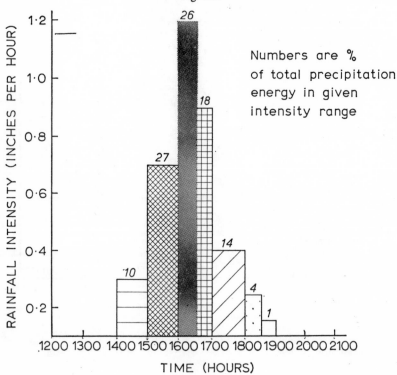

Numbers are % of total precipitation energy in given intensity range

Fig. 5.32

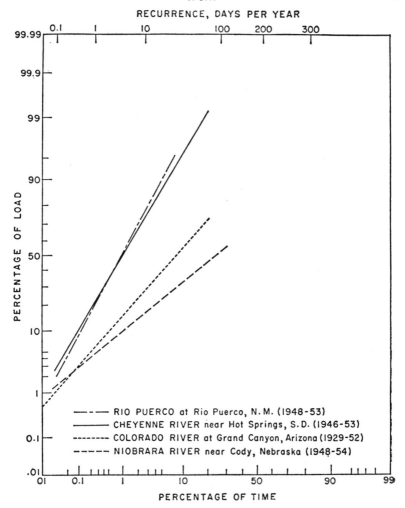

RECURRENCE, DAYS PER YEAR

PERCENTAGE OF TIME

----- RIO PUERCO at Rio Puerco, N.M. (1948-53)
——— CHEYENNE RIVER near Hot Springs, S.D. (1946-53)
------- COLORADO RIVER at Grand Canyon, Arizona (1929-52)
- - - - NIOBRARA RIVER near Cody, Nebraska (1948-54)

Fig. 5.33   Plot of cumulative percent of time against cumulative percentage of total suspended load transported by four United States' rivers. (From Wolman and Miller, 1960)

Fig. 5.31   Theoretical relations between magnitude of applied stress and (*a*) rate of debris movement, (*b*) frequency of stress occurrence and (*c*) product of rate and frequency (*a* and *b*). (From *Fluvial Processes in Geomorphology* by Luna B. Leopold, Gordon Wolman and John P. Miller. W. H. Freeman & Company. Copyright © 1964)

Fig. 5.32   Histogram of rainfall intensity as a function of time during a rainstorm. The percentage of total kinetic energy produced during the storm is shown for each intensity class by the figure at the top of each bar. (From *Fluvial Processes in Geomorphology* by Luna B. Leopold, Gordon Wolman and John P. Miller. W. H. Freeman & Company. Copyright © 1964)

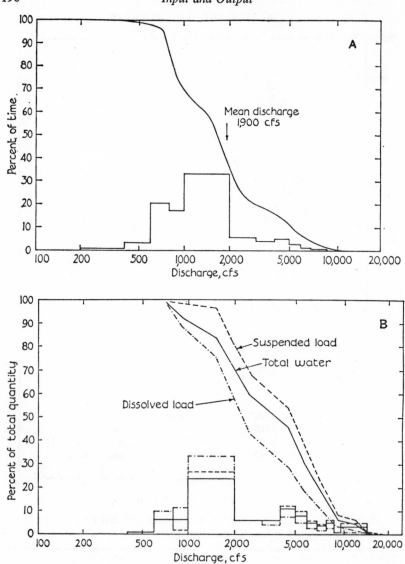

Fig. 5.34   Histograms and cumulative graphs of discharge and suspended and dissolved load for the Bighorn River at Thermopolis, Wyoming. (From *Fluvial Processes in Geomorphology* by Luna B. Leopold, Gorodon Wolman and John P. Miller. W. H. Freeman & Company. Copyright © 1964)

(*a*) Duration curve of discharge, showing percentage of time various discharge rates are equalled or exceeded.

(*b*) Cumulative graphs and histograms showing the relative contribution of various discharge conditions to total flow of water, suspended load and dissolved load.

thresholds, the crossing of which is accompanied by drastic increases of rates of change. The important question is to determine the magnitude of the energy events which are most effective in changing natural systems. It is apparent that, for example, in geomorphic systems the amount of work done depends on the frequency of energy events, as well as their magnitudes. Fig. 5.30 shows a typically logarithmic relationship between work done and event magnitude, and this type of relationship is common for most transporting processes, so that:

$$q = k(\tau - \tau_c)^n$$

where: q = rate of transport; τ = shear stress; τ_c = stress threshold; k = a constant relating to the material; and n = an exponent. It has already been shown that the frequency distributions of natural events are most commonly right-skewed and a graphical combination of these magnitude, frequency and work rate concepts (Fig. 5.31) has given rise to the notion that over a long period most work is accomplished by relatively frequent events of moderate intensity—the effects of events of high intensity being more than compensated for by their infrequent occurrence.

There is much empirical data to support the above hypothesis. The following figures give estimates of sand movement for Kharga Oasis, Egypt:

Wind velocity (m/s)	Sand movement (tons/year × 10⁴)
5·8–10	8·7
10–13·5	32·0
13·5–15·7	12·0

Fig. 5.32 shows the percentage of the total kinetic energy released during each time period of a typical moderate rainstorm of total 2·62 inches, suggesting the high cumulative effect of the lower intensity parts of the storm. Estimates of material moved by rivers support the efficacy of middle intensity processes in systems transformation. Fig. 5.33 gives the amount of suspended sediment moved by four rivers of the western U.S.A., indicating that 98–99% is carried during discharge events which have a return period of less than 10 years. Other estimates of 192 U.S. basins of less than 8 square miles give 50% of total debris moved by runoff of return periods less than 28 years. For larger basins and more complete records, however, it is believed that most of the debris removed is accomplished by runoffs of return periods of less than 2 years. An analysis of the relationship of load to discharge of the Bighorn River (Fig. 5.34) leads to the conclusion that discharges greater than 10,000 cfs transport only some 9% of the total suspended load, and discharge levels occurring on 9 out of 10 days (i.e. less than 5,500 cfs) transport about 57%. The most complete work of this sort has calculated the total material moved (suspended, dissolved and bed loads) in a 6-day period of

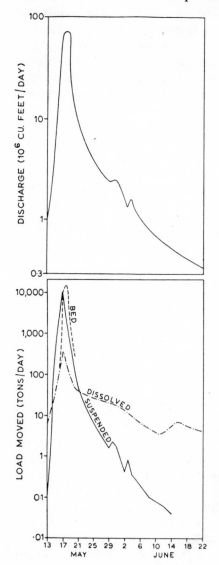

Fig. 5.35   Discharge, together with bed load, suspended load and dissolved load, associated with a 50-year flood in the Graburn Watershed, Alberta. (After McPherson and Rannie, 1969)

flood discharge in Graburn Watershed, Alberta (Fig. 5.35), of 1600 feet relief and 15·8 square miles in area. The effect of this at least 50-year flood was estimated at an equivalent lowering of the whole basin by 0·0013 feet, and this figure is only of the order of 5 times greater than the mean annual erosion rates estimated for U.S. basins of about the same size.

Although the above evidence seems to support the hypothesis of the relative efficiency of work performed by middle intensity processes, it is clear that the calculations are mostly based on individual isolated events of middle and low magnitude. There is little real information on the total work performed by infrequent high intensity events which follow closely upon one another, as in the 1952 River Lyn disaster in England. In such cases one event would vastly reinforce the effect of the other, but their very infrequency makes anything more than speculation impossible at present. Chapter 6 will deal with the evidence for equilibrium states in systems and in this discussion one must return to this question of the magnitude of energy events to which adjustment appears to be made.

REFERENCES

Anderson, M. G. (In preparation), Relation of the inter-arrival time distributions of precipitation and discharge within small drainage basins.

Barry, R. G. and Chorley, R. J. (1968), *Atmosphere, Weather and Climate* (Methuen, London), 319.

Benson, M. A. (1962), Evolution of methods for evaluating the occurrence of floods; *U.S. Geological Survey, Water Supply Paper* 1580-*A*, 30.

Benson, M. A. and Thomas, D. M. (1966), A definition of dominant discharge; *Bulletin of the International Association of Scientific Hydrology*, Year XI (1), 76–80.

Caine, N. (1968), The log-normal distribution and rates of soil movement: An example; *Revue de Geomorphologie Dynamique*, Year XVIII, No. 1, 1–7.

Caskey, J. E. (1963), A Markov chain model for the probability of precipitation occurrence in intervals of various length; *Monthly Weather Review*, 92, 298–301.

Chorafas, D. N. (1965), *Systems and Simulation* (Academic Press, New York), Chapter 15.

Chorley, R. J. (Ed.), (1969), *Water, Earth and Man* (Methuen, London), 588.

Chorley, R. J. (1971), Forecasting in the earth sciences; In Chisholm, M., Frey., A. E. and Haggett, P., (Eds). *Regional Forecasting* (Butterworths, London), 121–137.

Cole, G. (1966), The application of regional analysis of flood flows; *Institution of Civil Engineers, Symposium on River Flood Hydrology*, 39–57.

Conrad, V. and Pollak, L. W. (1950), *Methods in Climatology* (Harvard), 459.

Court, A. (1951), Temperature frequencies in the United States; *Journal of Meteorology*, 8, 367–380.

Court, A. (1952), Some new statistical techniques in geophysics; *Advances in Geophysics*, 1, 45–85.

Court, A. (1953), Temperature extremes in the United States; *Geographical Review*, 43, 40–49.

Curry, L. (1965), A stochastic model for soil moisture balance (Abst.); *Annals of the Association of American Geographers*, 55, 611.

Dalrymple, T. (1960), Flood frequency analyses; *U.S. Geological Survey Water Supply Paper A–1543*, 80.

Dury, G. H. (1964), Some results of a magnitude-frequency analysis of precipitation; *Australian Geographical Studies*, 2, 21–34.

Feyerherm, A. M. and Bark, L. D. (1965), Statistical methods for persistent precipitation patterns; *Journal of Applied Meteorology*, **4**, 320–328.

Gabriel, K. R. and Neumann, J. (1962), A Markov chain model for daily occurrence at Tel Aviv; *Quarterly Journal of the Royal Meteorological Society*, **88**, 90–95.

Gordon, G. (1969), *System Simulation;* (Prentice Hall Inc., New Jersey), 303.

Gregory, S. (1957), Annual rainfall probability maps of the British Isles; *Quarterly Journal of the Royal Meteorological Society*, **83**, 543–549.

Gregory, S. (1963), *Statistical Methods and the Geographer* (Longmans, London), 240.

Gumbel, E. J. (1941), Probability-interpretation of the observed return-periods of floods; *Transactions of the American Geophysical Union*, 836–849.

Gumbel, E. J. (1958A), *Statistics of Extremes* (Columbia Univ. Press, New York), 375.

Gumbel, E. J. (1958B), Statistical theory of floods and droughts; *Journal of the Institution of Water Engineers*, **12**, 157–184.

Haggett, P. and Chorley, R. J. (1969), *Network Analysis in Geography* (Arnold, London), 348.

Hershfield, D. M. and Kohler, M. A. (1960), An empirical appraisal of the Gumbel extreme-value procedure; *Journal of Geophysical Research*, **65**, 1737–1746.

Hershfield, D. M. (1962), A note on the variability of annual precipitation; *Journal of Applied Meteorology*, **1**, 575–578.

Kirby, W. (1969), On the random occurrence of major floods; *Water Resources Research*, **5**, 778–784.

Leopold, L. B. and Maddock, T. (1953), The hydraulic geometry of stream channels and some physiographic implications; *U.S. Geological Survey, Professional Paper 252*; 57.

Leopold, L. B., Wolman, M. G. and Miller, J. P. (1964), *Fluvial Processes in Geomorphology* (Freeman, San Francisco), 522.

McPherson, H. J. and Rannie, W. F. (1969), Geomorphic effects of the May 1967 flood in Graburn Watershed, Cypress Hills, Alberta, Canada; *Journal of Hydrology*, **9**, 307–321.

Putz, R. R. (1952), Statistical distributions for ocean waves; *Transactions of the American Geophysical Union*, **33**, 685–692.

Reich, B. M. (1963), Short-duration rainfall-intensity estimates and other design aids for regions of sparse data; *Journal of Hydrology*, **1**, 3–28.

Rodda, J. C. (1967), A country-wide study of intense rainfall for the United Kingdom; *Journal of Hydrology*, **5**, 58–69.

Stidd, C. K. (1953), Cube-root-normal precipitation distributions; *Transactions of the American Geophysical Union*, **34**, 31–35.

Strahler, A. N. (1965), *Introduction to Physical Geography* (Wiley, New York).

Todd, D. K. (1953), Stream-flow frequency distributions in California; *Transactions of the American Geophysical Union*, **34**, 897–905.

Wolman, M. G. and Miller, J. P. (1960), Magnitude and frequency of forces in geomorphic processes; *Journal of Geology*, **68**, 54–74.

6: *Systems Equilibrium*

> Nature . . . creates ever new forms; what exists
> has never existed before, what has existed returns
> not again—everything is new and yet always old. . . .
> There is an eternal life, a coming into being and
> a movement in her; and yet she goes not forward.
>
> GOETHE: *Essay on Nature*

6.1 The Importance of Equilibrium

Implicit in the foregoing discussion of the characteristics and operation of systems in physical geography has been the notion that feedback mechanisms are constantly at work, tending to bring about some balance, or equilibrium, between the morphological and cascade components. Equilibrium, however, is a highly ambiguous state which presents many different aspects and is the subject of a wide variety of definitions. The following are a few of the terms, sometimes conflicting, which are commonly used to attempt to define the various equilibrium states:

1 Static equilibrium—in which a balance of tendencies brings about a static condition of certain system properties, both absolutely and relatively (Fig. 6.1*a*). When velocities are balanced the condition is sometimes termed 'stationary equilibrium' and where a balance of opposing forces causes the resultant force to vanish the term 'poised equilibrium' is often applied. A clear example of this state, 'macroscopic equilibrium', occurs when, over a reasonable length of time, none of the observable macroscopic system properties of interest to the observer changes appreciably.

2 Stable equilibrium—being the tendency for a system to move back toward a previous equilibrium condition (i.e. to 'recover') after being disturbed by limited external forces (Fig. 6.1*b*).

3 Unstable equilibrium—wherein a small displacement leads to a greater displacement, usually terminated by the achievement of a new stable equilibrium (Fig. 6.1*c*).

4 Metastable equilibrium—when stable equilibrium obtains only in the absence of a suitable trigger, catalyst or minimum force, which carries the system state over some threshold into a new equilibrium regime (Fig. 6.1*d*).

201

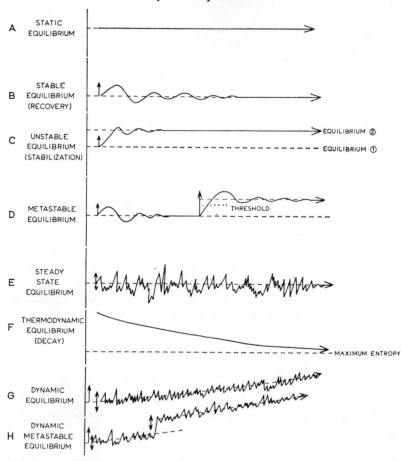

Fig. 6.1   Diagrammatic representation of eight equilibrium conditions.

5 Steady state equilibrium—a state of an open system wherein properties are invariant when considered with reference to a given time scale, but within which its instantaneous condition may oscillate due to the presence of interacting variables (Fig. 6.1*e*). Akin to this is 'statistical equilibrium', in which the frequencies of occurrence of component populations continue to remain proportionally allocated according to certain probabilities regardless of the vagaries of individual components, due to an absence of trends, cycles or any other timebound patterns in the processes. Thus the equilibrium of a Markov chain system in which individuals are constantly varying can be defined by the presence of certain stable probabilities wherein the proportions of individuals within classes remain constant.

6 Thermodynamic equilibrium—a tendency towards a condition of maximum *entropy* in an isolated system, expressed by the *Second Law of Thermodynamics*, that there is an increase of entropy in every natural process, providing all the system taking part in the process is considered (Fig. 6.1*f*). Maximum entropy is defined, in this context, as an equal probability of encountering given states, events or energy levels throughout the system, and is associated with an extinction of energy differences (i.e. free energy) capable of performing work within the system (in other words the energy distribution is the 'most probable'), and with maximum randomness (or disorder) and uniformity—as opposed to the organization and hierarchical structure exhibited by systems at low entropy.

7 Dynamic equilibrium—balanced fluctuations about a constantly changing system condition which has a trajectory of unrepeated 'average' states through time. The rate of change of the fluctuations is so much greater than that of the average system state that, when observed instantaneously or on a short time scale, the former masks the latter to give the appearance of a steady state equilibrium (Fig. 6.1*g*). This tendency towards a steady state, maintained in the face of a constantly changing gross energy environment so that absolute equilibrium is seldom actually present, is also termed 'quasi-equilibrium'. Where thresholds allow occasionally great fluctuations to initiate a new regime of dynamic equilibrium a more complex state of 'dynamic metastable equilibrium' exists (Fig. 6.1*h*).

From these definitions it is clear that a number of important concepts and reservations underlie the many aspects of equilibrium, and an elaboration of these concepts within the context of physical geography forms the framework for the remainder of this chapter. These include:

1 The existence of *self-regulation*, sometimes called negative feedback, homeostasis, stabilization, damping or recovery. It should be stressed that although negative feedback loops are usually dominant in the operation of physical systems, some other systems—notably those of which man plays a part—are direction-amplifying (i.e. *morphogenetic*). More will be said on this point in Chapter 8.

2 *Correlation* of properties of a system often, but not always, expresses a measure of its equilibrium state.

3 *Statistical stability*, arrived at by either mutual adjustments among the systems components or by progressive decay of free energy, is another possible measure of equilibrium. The implication is that equilibrium is associated with 'something' remaining constant, even if this constancy

15

exists for only a limited time period and can only be recognised statistically.

4 The idea of *optimum efficiency* of operation is commonly associated with equilibrium, and is expressed by such phrases as 'balance of forces', 'minimum work' and 'conservation of energy'.

5 *Balance* between opposing tendencies is so basic to the notion of equilibrium that the former term is sometimes used as a synonym for the latter. It should be stressed that equilibrium seldom, if ever, involves the relations of all the variables in a system, and that any equilibrium state must be defined with reference to the nature and scale of the variables concerned.

6 *Thresholds* exist across which recovery and self-regulation processes act with difficulty. Nevertheless these thresholds have equilibrium significance in that they separate different system regimes or economies within each of which characteristic equilibrium states are possible.

7 External change initiates different *relaxation paths* of system variables, depending both on the composition and state of the system, and on the nature and energy environment of the change.

8 There is often a complex interrelationship between *equilibrium tendencies* and those associated with *temporal change*, characterized by the definition of dynamic equilibrium. The equilibrium state can only be satisfactorily specified in such circumstances with strict reference to a given length or scale of time.

One problem commonly encountered in the discussion of systems equilibrium is the distinction which is often drawn between the thermodynamic definitions of equilibrium referred to isolated systems and the dynamic definitions considered more appropriate to closed or open physical systems. This distinction is often an academic one, however, for complex closed and open systems can only be analysed by assuming that they consist of a number of subsystems, displaying virtually isolated system characteristics in some significant respects. Indeed, some would argue that attempts to define equilibrium only have meaning in an isolated system context, and that one can measure the degree of success in explanation of the real world by the degree to which complex closed and open systems can be broken down into virtually isolated subsystems whose state can be assessed with reference to some postulated equilibrium condition. In other words, the level of sophistication reached in any study of physical reality lies in the success of attempts to describe the conditions of isolation for the systems of which it is composed. Thus the concept of equilibrium allows us to make very useful simplifying assumptions regarding the nature of complex systems, and in the final analysis it is only when a system exhibits some kind of equilibrium conditions that it can be recognised to exist as an entity at all.

6.2 Self-Regulation

Self-regulation, or negative feedback, operates to counteract or damp down the effects of external changes on a system so that it either returns to an original equilibrium condition or stabilizes itself in a new equilibrium. This tendency to maintain a certain constancy of operation in the face of external fluctuations is termed *homeostasis* when applied to biological systems, and is characterized, for example, by the involuntary regulation of body temperature and blood pressure in mammals. Negative feedback is achieved by certain components of a process-response system operating to activate tendencies which oppose changes in external inputs and which are proportional in magnitude to these input changes. As has been pointed out animate and inanimate systems may behave very differently in the presence of changing inputs, the latter always, in the long run, tending towards stability, whereas the former need not necessarily be dominantly 'teleological' or 'goal seeking' in respect of some preferred state. Animate systems at a scale above that of the individual organism commonly display dominantly positive feedback or 'deviation-amplifying' (morphogenetic) features wherein a vicious circle allows, for example, a snowballing effect of capital accumulation and the increasing differentiation of rich and poor. In all types of system, negative feedback is accomplished by one, or both, of two types of mechanism:

1 The *servomechanism*, forming a loop in the system response so that some counteracting restraint is placed on changes of external input.
2 The operation of some system components as *regulators* which, as their own characteristics are modified, tend to stabilize the system internally in a steady state in the presence of external energy changes which the system must accept.

Cybernetics, the 'science of control' (from the Greek word meaning 'steersman'), has been much concerned with the role of servomechanisms which can be considered as tracking systems monitoring the corrections of energy input and causing the system to exert an input control proportional, and in the opposite sense, to these variations. Fig. 6.2 shows a simple servomechanism involving negative feedback where the output from the process (y), which has operated on the inputs through a transfer function (K), is measured by a sensor such that some proportion (b) of it is fed back to the input (x) by means of a comparator which changes the input in a compensating direction

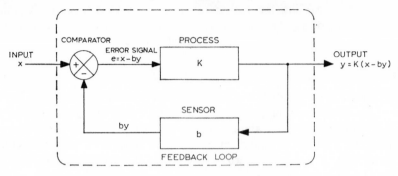

Fig. 6.2   A simple servomechanism.

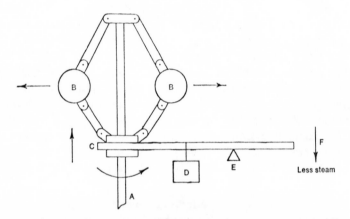

Fig. 6.3   The Watt ball governor. The engine rotates the shaft (A) which causes the balls (B) to spread, raising the floating collar (C) and the weight (D). This rise operates via the fixed pivot (E) to close a valve (F) which decreases the stream supply to the engine. Thus the engine can be governed to maintain a more-or-less constant rate of operation which is pre-determined by adjusting the size and position of the weight (D). (From Hare, 1967)

to produce an error signal (*e*) whereby the input is modified. The Watt ball governor (Fig. 6.3) is the classic example of a mechanical negative feedback device. Unfortunately, such a simple concept has little important application in physical geography largely because few of the systems can exercise a decisive control over their inputs. An example of such a control might be the initial decrease in infiltration capacity of a soil system as the downward movement of the wetting front is impeded by trapped soil air before the whole soil column is wetted and unimpeded infiltration can commence. Other reasons for the limitations of the above concept are that most natural systems:

1 Have many integrated feedback mechanisms, the variety of which gives considerable flexibility of response by the system to external changes.
2 Exhibit a time-delay, or lag, between changes of input and the time when corrective action is applied. The operation of systems with lag having a wide range of input variations is particularly complex when the time-scale of the lag is long compared with the time intervals between possible input changes.
3 Show highly unstable outputs where the pattern of input changes into a lagging system involves a time trajectory.

The modification of system parameters whereby regulation of a system is achieved is much the most important method of self-regulation possessed by physical systems. This tendency of internal adjustment so as to counteract the effects of external changes is generally very difficult to analyse in all but the grossest manner. Such a tendency operates either to reinstate an original equilibrium condition by applying a counteracting rate of change initially proportional to the degree of disequilibrium introduced (thereafter decreasing according to some negative exponential function), or to produce a new dynamic equilibrium by distributing the effects of the external disturbance among the appropriate constituent parts of the system so that a minimum change in any one of them is involved. This tendency is stated in the context of physical chemistry by *Le Chatelier's principle* such that, if a change of stress is brought to bear on a system in equilibrium, producing a change in any of its components, a reaction occurs so as to displace the system state in a direction which tends to absorb the effect of the change, such that, if this reaction occurred alone, it would produce a change in an opposite sense to that of the original stress change. A classic example of the operation of such feedback in geomorphology occurs if a bar of coarse material is locally emplaced in a stream channel by bank caving or an isolated peak discharge. The bar will be associated with a steepening of the bed gradient, an increase of velocity and erosion rate, and a tendency for a return to the original bed gradient. A further example would involve an extension of the soil infiltration capacity mechanism when, after a period of high initial infiltration capacity (F^*), packing, the swelling of soil particles, and the clogging of pore spaces by fine particles reduces the infiltration capacity to a limiting equilibrium value (f_c) characteristic of the given soil (see Fig. 3.32). When two systems are linked together they may mutually exert negative feedback on each other, as when an increase of downcutting in a broad stream channel leads to a steepening of the associated valley-side slopes which, in turn, causes an

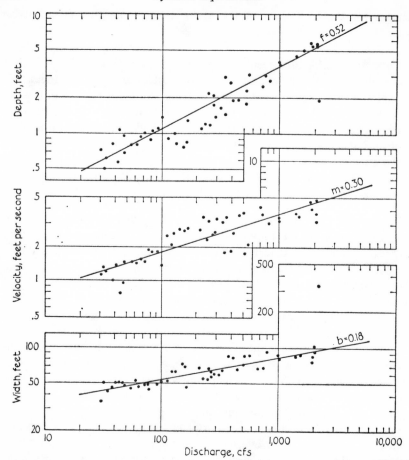

Fig. 6.4   At-a-station changes of width, mean depth and mean velocity with discharge of Seneca Creek at Dawsonville, Maryland. (From *Fluvial Processes in Geomorphology* by Luna B. Leopold, Gordon Wolman and John P. Miller. W. H. Freeman & Company. Copyright © 1964)

increased sediment yield to the channel so counteracting the downcutting.

Besides the complexities noted previously, the effect of negative feedback mechanisms is further complicated by the existence of:

1 Secondary responses. These are longer term results of the primary responses as when, for example, a change in rainfall may change the discharge of a stream setting in train primary feedback mechanisms producing channel stability, only to find that subsequent vegetational changes in the catchment resulting from the same change in rainfall may further alter the discharge and the hydraulic geometry of the channel.

2 Thresholds. The passage of the system through one of these may involve drastic and perhaps non-recoverable changes. An example of this is when an increase of runoff is sufficient to strip off the surface vegetation and initiate a system of gullies which cannot be obliterated even after a prolonged return to the original runoff conditions.

6.3 Correlation of Properties

One of the most common methods of recognising the existence of some kind of equilibrium condition is through the identification of significant correlations between variables within systems. For example, the plot of discharge at-a-station against suspended sediment load (see Fig. 5.30) implies a close response relationship between a process and an aspect of the work accomplished on suitable available channel material. Such work is also reflected in the almost instantaneous adjustments between at-a-station discharge (Q) on the one hand, and stream depth (D) and width (W), as well as mean velocity (V), on the other (Fig. 6.4).

In this instance:

$$W \propto Q^b$$
$$D \propto Q^f$$
$$V \propto Q^m$$

also

$$n \propto Q^y$$

and

$$S \propto Q^z$$

where n = roughness and S = the slope of the energy line

$$WDV = Q$$

$$\therefore Q^b \cdot Q^f \cdot Q^m \propto Q^1$$

$\therefore$ In theory $b + f + m = 1$ (this is not always exactly so in practice).

These relationships are such that any variation in discharge causes changes in velocity and shear stress which are translated into changes in width and depth (or, more precisely, hydraulic radius) at-a-station, each of which does not carry equal weight, and the relationships between the adjustments differ from one station to another. Similar equilibrium relationships have been shown to hold between the above variables at stations along a river, as well as between discharge and water surface slope and channel roughness (n') (Fig. 6.5). Channel roughness (n'), calculated by substituting the remaining values in the modified Manning formula ($V = (1 \cdot 486/n') \, D^{2/3} \, S^{\frac{1}{2}}$), is a measure

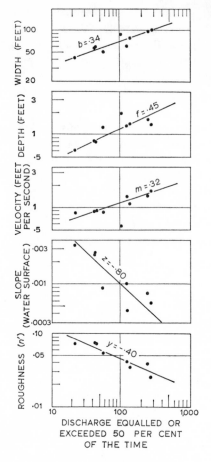

Fig. 6.5

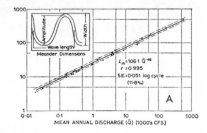
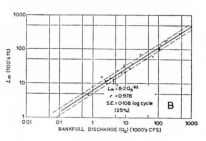

Fig. 6.6

Fig. 6.5 Relations of width, depth, velocity, water surface slope and channel roughness to discharges equalled or exceeded 50 per cent of the time for stations along Brandywine Creek, Pennsylvania. (From Wolman, 1955)

Fig. 6.6 Relations of meander wavelength to (*a*) mean annual discharge and to (*b*) bankfull discharge for a number of rivers in the eastern United States. (After Carlston. From Haggett and Chorley, 1969)

of both the micro- and macro-roughness of the channel. These relationships imply that changes in channel geometry are adjusted to the independent variables of discharge and load. A similar equilibrium relationship involving process and form has been commonly proposed between a characteristic channel forming discharge (i.e. bankful (Q_b), or mean annual flood ($\bar{Q}$)) and the associated meander wave length (Fig. 6.6).

Another aspect of correlation, held to be indicative of equilibrium conditions, is when a high correlation structure of morphological variables is observed. Fig. 6.7 shows such a high interlocking of variables relating to valley-side slope and associated basal stream systems on the Charmouthien Limestone of Eastern France where the streams are actively working at the

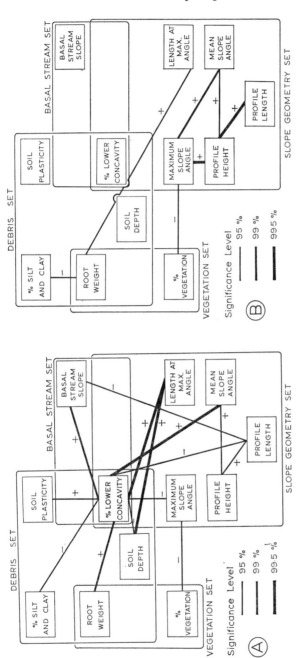

Fig. 6.7 The interlocking of variables relating to slope geometry, debris, vegetation and basal stream activity for slopes on the Charmouthien Limestone, Plateau de Bassigny, Northern France.

(a) Where the stream is adjacent to the slope base.
(b) Where the stream is away from the slope base.
(After Kennedy, 1965. From Chorley and Haggett, 1967)

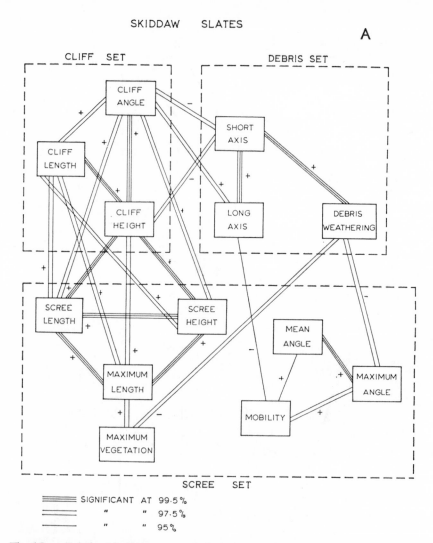

Fig. 6.8   Relationships between morphological properties of cliff and scree slopes in the English Lake District. (After Towler, 1969)

(a) The Skiddaw Slates.
(b) The Borrowdale Volcanics.
(c) A generalized diagram defining the morphological properties.

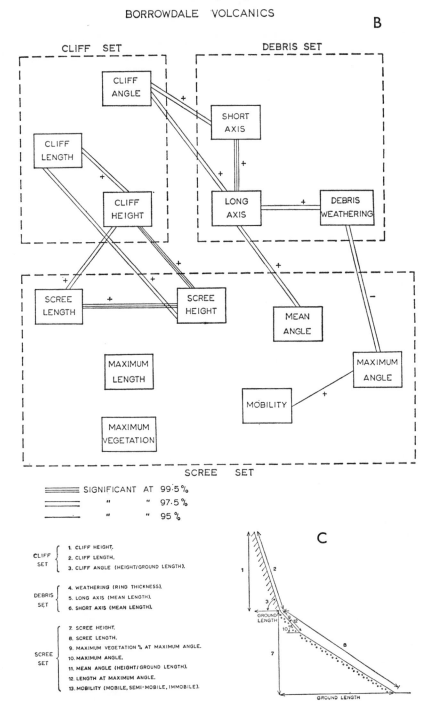

BORROWDALE VOLCANICS

B

CLIFF SET   DEBRIS SET

CLIFF ANGLE

SHORT AXIS

CLIFF LENGTH

CLIFF HEIGHT

LONG AXIS   DEBRIS WEATHERING

SCREE LENGTH   SCREE HEIGHT

MEAN ANGLE

MAXIMUM LENGTH

MAXIMUM ANGLE

MOBILITY

MAXIMUM VEGETATION

SCREE SET

═══════ SIGNIFICANT AT 99.5%
─────── "   " 97.5%
───── · ── "   " 95%

C

CLIFF SET
{
1. CLIFF HEIGHT.
2. CLIFF LENGTH.
3. CLIFF ANGLE (HEIGHT/GROUND LENGTH).
}

DEBRIS SET
{
4. WEATHERING (RIND THICKNESS).
5. LONG AXIS (MEAN LENGTH).
6. SHORT AXIS (MEAN LENGTH).
}

SCREE SET
{
7. SCREE HEIGHT.
8. SCREE LENGTH.
9. MAXIMUM VEGETATION % AT MAXIMUM ANGLE.
10. MAXIMUM ANGLE.
11. MEAN ANGLE (HEIGHT/GROUND LENGTH).
12. LENGTH AT MAXIMUM ANGLE.
13. MOBILITY (MOBILE, SEMI-MOBILE, IMMOBILE).
}

GROUND LENGTH

GROUND LENGTH

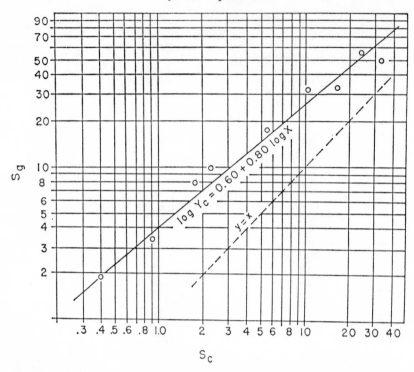

Fig. 6.9　　Regional plot of mean valley-side slope (S_g) against mean channel gradient (S_c) for nine maturely-dissected regions. (After Strahler, 1950)

slope base, as well as the general paucity of significant correlations in locations where basal stream activity is lacking. The implication of such an observation is that where stream and slope processes are operating in conjunction at present this promotes some kind of equilibrium of operation within the system which is manifested by high correlations between all sorts of morphological parameters of stream and slope geometry, as well as debris and vegetation parameters. In short, the former case seems indicative of some kind of functional interaction, whereas the latter does not. From this viewpoint any present landscape appears more and more as a complex of interlocking systems exhibiting a whole spectrum of states from instantaneous equilibrium to complete disequilibrium. A similar analysis of the cliff and basal scree systems of the Skiddaw Slates and Borrowdale Volcanies in the British Lake District (Fig. 6.8) strongly indicates that some equilibrium conditions exist between the geometrical features of the cliffs, basal scree slopes and component debris of the Slates, which are absent from the similar features of the Volcanics. The natural extension of this argument is that

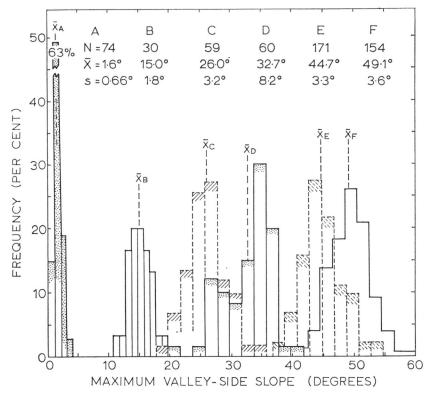

Fig. 6.10   Histograms of maximum valley-side slope angle frequencies from six regions.
(From Strahler, 1950):

- (*a*) Steenvoorde, France.
- (*b*) Rose Well gravels, Arizona.
- (*c*) Bernalillo, New Mexico (Santa Fe formation).
- (*d*) Hunter-Shandaken area, Catskill Mountains, New York.
- (*e*) Kline Canyon area, Verdugo Hills, California.
- (*f*) Dissected clay fill, Perth Amboy badlands, New Jersey.

forms are more related to contemporary processes in the highly-correlated systems, whereas the more poorly-correlated are of largely relict systems on which the present processes are operating haphazardly and much less efficiently than previous processes. Of course, one has to exercise some care in drawing conclusions regarding the existence of a contemporary equilibrium state from observed statistical correlations in that some high correlations result from the operation of mutual constraints (i.e. such auto-correlation as that which tends to exist geometrically between slope length and slope height), and the remains of morphological adjustments to past processes would show similar high correlations simply because subsequent processes

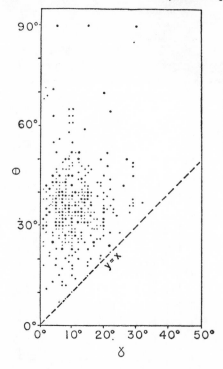

Fig. 6.11   Relation between 433 individual channel gradients (γ) and the associated maximum valley-side slope angles (θ) in Lighthouse Hollow, a tributary of the Farmington Rover, Connecticut. (From Carter and Chorley, 1961.) First-order basin data have been omitted. Large dots indicate more than one coincident reading.

have been operating too weakly or for too short a period to change the former morphological relationships. This argument, however, applies much less forcibly to variables relating to soil and vegetation characteristics, which have a shorter relaxation time.

Equilibrium relationships are often inferred from the existence of high correlations between generalized or averaged measures. For example averaged regional maximum valley-side slopes (S_g) and the slopes of the associated basal channels (S_c) (Fig. 6.9) seem to exhibit some equilibrium adjustment between the geometry of individual slopes and that of their basal channels so as to allow the steady removal by the latter of debris supplied by the former. The difference of the slope ratio S_c/S_g being perhaps due to differing climatic and vegetative conditions among the nine regions. These means, however, mask wide ranges of individual slope geometries within each region (Fig. 6.10), such that plots of θ versus γ for each measured slope and channel within an area do not necessarily show the above simple relationship (Fig. 6.11). There is also some evidence (Fig. 6.12) to suggest that this relationship can differ very markedly between areas. This variation in the form of association between S_c and S_g is due to the fact that streams may

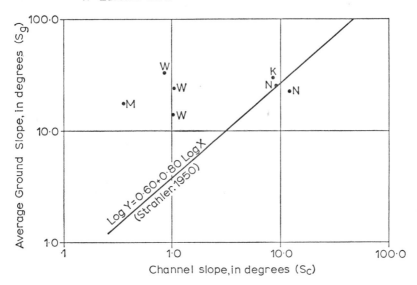

Fig. 6.12   Differences in the relationship between average channel slope (S_c) and average ground slope (S_g) between areas. (From Kennedy, 1969)

adjust to the erosional environment of adjacent slopes by changes in *cross-section* as well as in long profile. The equilibrium significance of such types of statistical stability is treated in the next section.

6.4 Statistical Stability

One of the classical indicators of equilibrium in physical geography is the cancelling-out of the effects of events which seem randomly distributed in time about some average value. This statistical stability of a time series, which is embodied in the notion of equality in cut and fill of a 'graded' stream, is illustrated diagrammatically in Fig. 6.13*a* and contrasted with three other types of series. Fig. 6.13*b* and *c* illustrate series having strong timebound components which are treated in Chapter 7, and the occurrence of an isolated high deviation in Fig. 6.13*d* introduces the possibility that a threshold may be transgressed. Statistical stability of a time series is implied when rapid scour and fill of a channel allow its form and gradient to remain to all intents and purposes in a statistically constant quasi-equilibrium with the available discharge and sediment characteristics, even while the channel is slowly eating away the land. Again, the 'equilibrium beach profile' exhibits

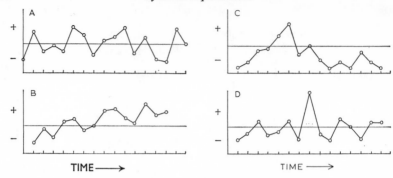

Fig. 6.13   Time-series of observations illustrating statistical stability and instability. (From Hare, 1967)

> (*a*) Random fluctuation, implying statistical stability.
> (*b*) The presence of a time trend.
> (*c*) A possible periodic pattern.
> (*d*) A possibly otherwise stable series containing one highly improbable deviation.

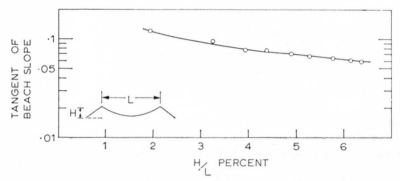

Fig. 6.14   Relation of beach slope to wave steepness for beaches along the West Coast of the United States having beach material of 0·4 mm calibre. (After Wolman, and Miller, 1960)

a similar tendency when, despite rapid daily and seasonal fluctuations, there is over a period of years a recognisable profile by which a given beach may be characterized. Beach profiles are dominantly controlled by wave steepness (H/L—an inverse control) and by debris calibre (directly), and Fig. 6.14 shows the relationship between H/L and beach slopes along the west coast of the U.S.A. for beach calibres of 0·4 mm. Ordinary waves (i.e. with $H/L < 2·5\%$) vary so that the short, choppy winter waves steepen the beach, whereas the longer waves of summer flatten the profile. The destruction of these relationships by occasional storm conditions ($H/L > 2·5\%$) forms a threshold which can be recrossed (i.e. recovered) on the return to more ordinary wave conditions.

When the concept of equilibrium as manifested by statistical central tendencies or clustering is transferred from a temporal into a spatial context, one arrives at the notion that equilibrium states may be recognised by the tendency for rather close clustering of geometrical terrain variables around certain characteristic values (see Fig. 6.10). Viewed in such a light landform regularity in a given region represents the result of a series of constraints due to an equilibrium distribution of hydraulic energy over the land surface, with deviations from uniform conditions resulting both from chance local variations (e.g. in lithology) and from deviations about mean conditions initiated by external changes. The forms will tend to lag behind these external changes, as does the operation of a sluggish governor or an old steam engine. These lags may sometimes be of very short duration, however, and the fluctuations of hydraulic geometry are often measurable in the hours, or even minutes, during which changes of discharge occur. Of course, the two sources of variance are largely indistinguishable in practice. Equilibrium in a landscape manifests itself in a tendency towards a mean condition of unit forms, recognisable statistically, about which variations may take place over periods of time associated either with fluctuations in the energy flow (steady state) or the progressive change in the system condition (timebound changes), or both (dynamic equilibrium). Under such conditions, therefore, in areas of essentially uniform lithology, soils, vegetation, climate and stage of dissection, corresponding morphological variables will, as regard their form and magnitude, tend to cluster around characteristic mean values. This is what gives the geomorphic region its aspects of uniformity. Examples of the clustering are shown in Fig. 6.15 for maximum slope angles of basally-protected and corraded slopes in the Verdugo Hills, California.

We have already mentioned the concept of progressive decay in isolated systems, leading towards equilibrium states of uniform randomness and disorganization (i.e. of maximum entropy). From such a standpoint the equilibrium operation of a complex system can be approached by methods akin to those of statistical mechanics, wherein our inability to understand a complex of interrelated processes forces us to concentrate on the statistically average description of apparently individually random events and occurrences. Although the concept of entropy was originally used to describe the distribution of energy in isolated thermodynamic systems (i.e. its differential distribution) and thus is a measure of a system's incapacity to convert thermal energy into mechanical work, the application of entropy has been recently widened so that it is now applied to define levels of information content and to describe the state of closed and open systems.

Because information may be regarded as a measure of the amount of organization, an increase of entropy can be equated with a decrease in organization and an increase in the probability that all system events or

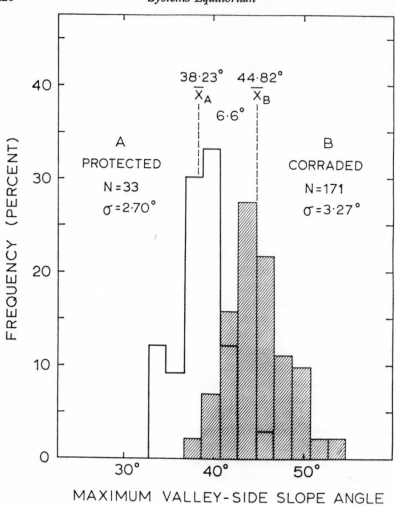

Fig. 6.15   Maximum valley-side slope angles of the Verdugo Hills, California, which
are (*a*) protected by basal debris accumulations, and (*b*) subject to basal stream corrasion.
(After Strahler, 1950)

states may be equally encountered either in time or space. Thus the information theory definition of entropy has a meaning independent of thermodynamics and is a synonym for 'probability' (thus being expressive of the
indeterminacy of the states of a system). If n events occur with the probability
$p_1, p_2, p_3 \ldots p_n$, where $\sum\limits_{i=1}^{n} p_i = 1$ (i.e. some of the given events are always

Fig. 6.16   Schematic diagrams of the 42 topologically-distinct channel networks with 11 links and 6 first-order Strahler streams. In a topologically-random population these networks would all be equally likely. The top row shows the possible second-order networks and the bottom four rows show the possible third-order networks. Arrowhead indicates outlet in each diagram. (After Shreve, 1966. From Haggett and Chorley, 1969)

bound to occur), then the entropy (ø) of the information system is given by the expression:

$$\emptyset = -\sum_{i=1}^{n} p_i \log_a p_i$$

Where $p_i = p_2 = p_3 = \ldots p_n$ (i.e. $p = 1/n$, and therefore any event or state is equally probable) ø is a maximum. ø can vary between zero, when there is complete certainty of one event always occurring, and unity, when there is complete indeterminacy in the system condition and no information is available. The total amount of information contained in a system is $\log_a n$, and if $\log_2$ is used (i.e. $a = 2$, rather than 10 or e) then the smallest possible unit of information is termed a 'bit'.

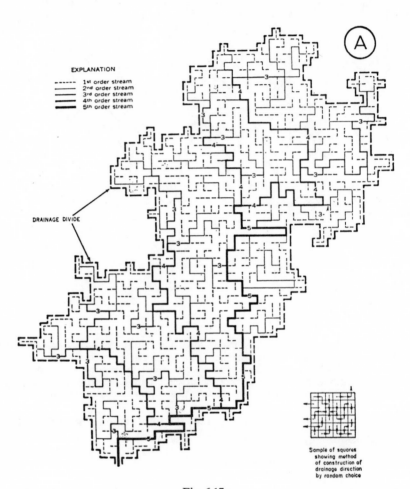

Fig. 6.17*a*

Fig. 6.17 Development of a random-walk drainage basin network (*a*), showing the relation of number and average length of streams to stream order (*b*). (From *Fluvial Processes in Geomorphology* by Luna B. Leopold, Gordon Wolman and John P. Miller. W. H. Freeman & Company. Copyright © 1964)

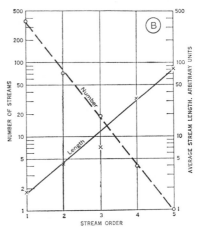

Fig. 6.17*b*

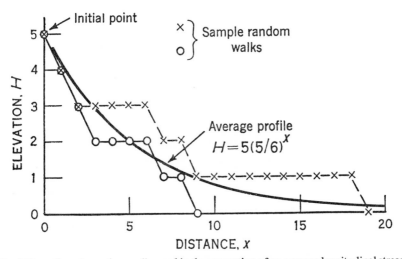

Fig. 6.18 Sample random walks used in the generation of an average longitudinal stream profile. (From *Fluvial Processes in Geomorphology* by Luna B. Leopold, Gordon Wolman and John P. Miller. W. H. Freeman & Company. Copyright © 1964)

One application of this concept of randomness or statistical indeterminacy to the notion of equilibrium states in physical geography has been in terms of interpreting areal stream patterns as having been produced by randomly-merging stream channels. If this is so then the combinations of patterns as produced make the occurrence of some sets of Strahler stream order numbers more probable than others. Fig. 6.16 shows, for example, all such combinations having 6 first order streams, 11 links and 5 forks, making up a topologically-distinct channel network family. A fourth-order basin having 27 first-order streams has statistically only 30 different possible sets of stream order numbers, but each set contains a different number of members (i.e. ways of combining the streams to produce the same order numbers) and therefore has a different probability of occurring randomly: viz.

RANK	ORDERS				PROBABILITY OF OCCURRENCE	
	1	2	3	4		
1	27	7	2	1	0·263	(gives slightly concave-up semi-logarithmic plot)
2	27	6	2	1	0·198	
3	27	8	2	1	0·169	
8	27	9	3	1	0·045	(follows Horton's law of stream numbers)
30						
					$\Sigma = 1·000$	

A further example of how the most likely averages developed from random occurrences are held to represent equilibrium or most likely states is given in Fig. 6.17. This shows a random-walk stream network, simulated by allowing drainage from each square with equal probability in each of the four cardinal directions, producing a stream network (*a*) in which the numbers of streams of each order closely approximate the laws of stream numbers and lengths (*b*).

The concept of entropy has also been more widely applied recently to open systems, where the import of both mass and energy occur. Because this import introduces differentiation, organization, order, free energy and information, it has been equated with the introduction of 'negative entropy' (i.e. *negentropy*). The imports of negentropy are regarded as continually changing the target maximum entropy to which the system is constantly tending to revert and thereby setting back its attainment of equilibrium. A move away from equilibrium is thus equated with a consumption of energy by the system and a decrease in its entropy. An open system in a steady state can therefore be defined as one in which the rate of increase of entropy is zero.

Entropy considerations have been used to rationalise the longitudinal profiles of rivers. Precipitation introduces inputs of potential energy at various elevations (H) of the stream system above base level (H_0), and the downstream movement of water and debris dissipates this, first as kinetic energy and ultimately as heat conduction, convection and radiation. If this system is in equilibrium (i.e. the internal increase of entropy is zero) then the rate of internal generation of entropy (ø) per unit mass discharge (Q) rate (i.e. $dø/dt \cdot 1/Q$) equals the rate of outflow of entropy (i.e. heat dissipation). If height above base level (i.e. potential energy) is considered as analogous to the absolute temperature in a thermodynamic system, the rate of energy loss per unit distance (x) along the channel is $dH/dx \cdot 1/H$. This expression represents the most probable distribution of energy losses in successive units of river length (dx) corresponding to a uniform rate of internal generation of entropy per unit length. Because a river is 'hydrologically indeterminate', it is free to adjust its depth, width, velocity or roughness to a given slope in many different ways to establish a longitudinal profile and hydraulic geometry wherein expenditure of work is minimized. The above theoretical expression for energy loss can be considered as a statement of the most efficient and probable longitudinal stream profile. This profile can be simulated by a series of simple random walks (Fig. 6.18) wherein, from an initial point above baselevel, the profile of flow is free to move in unit steps with two choices at each step:

p-downward one unit; where the value of p is made proportional to the corresponding height above baselevel;

q-continues on the same level for one unit; where the value of q is adjusted to changes in p so that $p + q = 1$.

Fig. 6.18 shows two simulated random walk profiles and the theoretical curve defining the mean position of all possible random walks generated by this simple model. This profile is exponential in form and similar to many natural river profiles.

6.5 Optimum Efficiency

The idea that natural processes constantly seek economy or maximum efficiency in all their operations is one of the oldest principles of theoretical science. Expressions of this maximum efficiency often occur as the result of the optimum design of each system component so that they work together most economically. Equilibrium, in this case, becomes an expression of optimality of working operation.

This 'optimality' concept is most often applied in respect of cascading systems and is usually linked with the notion of the equalization of work in space and time (as was discussed in the previous section) by the elimination of discontinuities. It is held, for example, that all open channel flow involving a

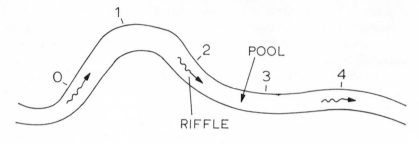

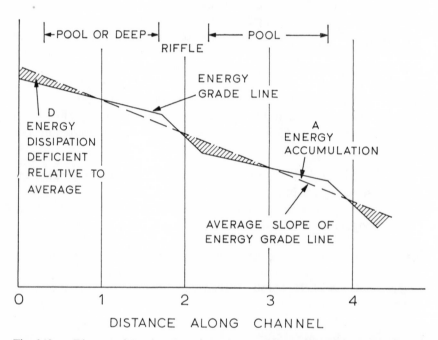

Fig. 6.19   Diagram of the plan view of a hypothetical reach of channel showing location of pools and riffles, and longitudinal profile of total energy (energy grade line) for the same reach. (From *Fluvial Processes in Geomorphology* by Luna B. Leopold, Gordon Wolman and John P. Miller. W. H. Freeman & Company. Copyright © 1964)

wide range of bed material sizes tends to produce accumulations of coarse debris (bars or 'riffles') along the bed at a spacing of 5–7 stream widths. This variation of micro-roughness results in a very unequal dissipation of energy along the stream, as evidenced by inequalities in the energy grade line (or water slope) (Fig. 6.19), which are especially apparent on straight streams. Equalization of energy dissipation is then restored by increasing the macro-roughness of the channel (i.e. bending it and dissipating energy by the interaction of turbulent flow paths) where the micro-roughness is least i.e. over the pools, such that on meandering streams the inequalities of energy dissipation are 'drowned out' at discharges greater than $Q_{2/3b}$ (i.e. two-thirds the bankfull discharge). This theory of the formation of meanders, if true, can only be a partial one because meanders are also found where there is bed debris of uniform calibre. It is important to recognise the importance of the concept of *equifinality* in natural phenomena—i.e. of similar end results emanating from different processes or historical sequences.

The classic notion of grade contains a strong element of the idea of optimal continuity of work, in which a graded stream profile is viewed as consisting of a system of segments each having that slope necessary to provide exactly the velocity required to transport the load supplied from upstream. The concave-up longitudinal profile is considered to result from the decrease in slope due to net downstream changes in discharge, load/discharge ratio, debris calibre, etc; but as none of these changes is systematic the resulting profile curve is not a simple mathematical one. This view of the graded (or equilibrium) profile has been criticised for its overemphasis on bed slope as providing the basic adjustment (negative feedback) mechanism and there may be reason to believe that adjustment of channel shape may be as significant a control over the continuity of transportation as that of the longitudinal profile. There is generally no way to predict whether a change in discharge or load charac-teristics can be more efficiently absorbed by a change in channel slope, rather than by a change in any, or all, of the other hydraulic variables (i.e. width, depth, macro-roughness, etc.). However, one of the definitions of equilibrium of a cascading system is that outputs of energy and/or material balance inputs in such a way so as to maintain not only constant levels of *integration* but also *certain peculiar gradients*.

6.6 Balance

The idea of balance is so closely associated with that of equilibrium that it needs to be stressed that often it is only certain of the system variables which are obviously involved in the balance and that there are a number of ways of defining or recognising its existence. Firstly, there is the balance which manifests itself in a more-or-less static condition reached as one major variable converges on some limit; secondly, there is the balance which is exhibited by the maintenance of a constant ratio between otherwise changing values; and, thirdly, there is the balance which can be shown to exist between clearly opposed forces or tendencies.

Horton's concept of the development of drainage channels to a limiting drainage density provides an example of the first type of balance, wherein such channels are considered to be exclusively the work of surface runoff

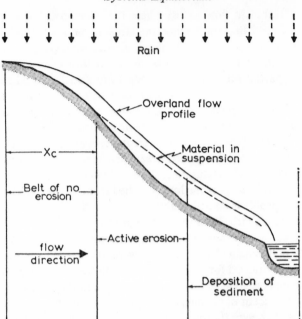

Fig. 6.20　　Hillslope profile illustrating surface erosion and flow processes according to the Horton model. (After Horton, 1945. From Kirkby and Chorley, 1967)

resulting from overland flow. The depth of this overland flow theoretically increases linearly away from any divide until, at a critical distance (X_c) (Fig. 6.20), it is deep enough to generate a velocity sufficient to entrain surface material and thereby initiate rills which coalesce by cross-grading to produce streams. The incision of these streams produces new valley-side slopes on which first rills and then new streams develop. The available length of overland flow (l_0) progressively decreases as these new slopes develop and this change is more rapid than the decline in X_c, because of the steepening of the slope profiles. The process continues until the drainage network has expanded to the point where the maximum available length of overland flow is reduced to a value characteristic of (i.e. adjusted to) the local conditions of geology, climate and vegetation. This minimum equilibrium value of X_c must be sufficiently small that not even the most intense runoff can initiate new rills upon the valley-side slopes, unless and until some major change occurs within the system (Fig. 6.21). This theory appears to be most applicable to arid badland conditions, whereas the presence of deep soil and vegetation complicates this simple view of the sequence of channel initiation occurring on the surface of a slope.

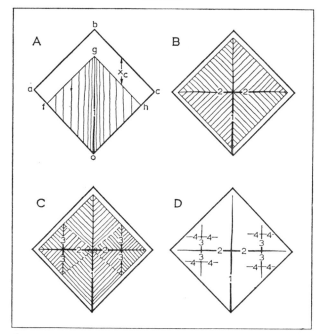

Fig. 6.21   Horton's model of the development of a drainage net
in a stream basin. (From Horton, 1945)

The idea that equilibrium manifests itself through the maintenance of characteristic ratios between dynamic elements is both highly ambiguous and difficult to substantiate. It has been suggested, for example, that in a state of erosional dynamic equilibrium an area of varied geology will have its geometry of relief and slope forms so mutually adjusted that all the lithological elements are downwearing at the same rate, implying a constant transformation without relative change in which all parts may be viewed as being adjusted to present processes. Thus for a quartzite to be comminuted and transported away at the same rate as an adjacent shale outcrop greater energy is needed so that the former requires and develops greater relief and steeper slopes than the latter. It has also been observationally demonstrated that in actively eroding small badland basins after about 25% of the theoretical mass of the basin has been removed (i.e. hypsometric integral $<75\%$) the basin relief ratios (H/L) and mean stream gradients ($\bar{\gamma}$) remain essentially constant until at least 80% has been removed (Fig. 6.22). Similarly, it appears that once the high point of a drainage basin is involved in the general degradation, at a hypsometric integral (see Fig. 2.18) of 60% or less, the integral more-or-less stabilizes itself, irrespective of the absolute relief, as the basin is geometrically transformed. Fig. 6.23 shows 5 regional hypsometric integrals, each averaged

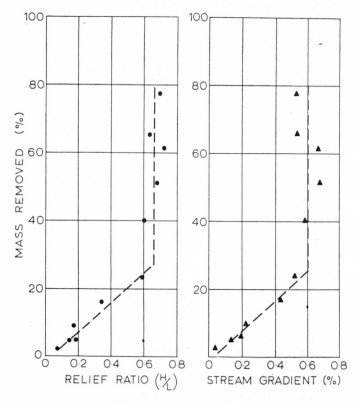

Fig. 6.22   Relations between percentage basin mass removed and (left) the relief ratio, and (right) the main stream gradient for erosional drainage basins in the Perth Amboy badlands, New Jersey. (After Schumm, 1956)

from 6 third or fourth order basins, representatives of which are depicted in Fig. 6.24. Thus within a comparatively small range of hypsometric integrals lies a wide range of basin relief, implying that, when applied to a single basin whose relief is being decreased through time, there is a long equilibrium (or mature) stage characterized by a more-or-less stable ratio of the hypsometric integral.

The above examples of the manifestation of equilibrium through the maintenance of characteristic ratios have been by-and-large empirical ones, but a theoretical principle from biology—that of *allometric growth*—also implies the maintenance of such ratios. The allometric law relates different parts of an organic system in dynamic equilibrium such that, as the system as a whole grows, the ratios between each part and the whole (and consequently between the parts) remain constant. This is normally stated as—the specific

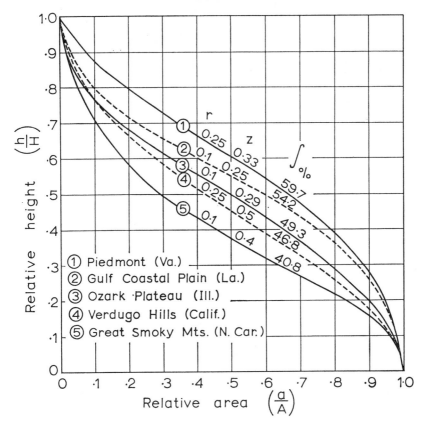

Fig. 6.23   Mean hypsometric curves of five areas in the equilibrium stage:
(1) Belmont Quadrangle, Virginia, U.S.A. N.S., 1:25,000.
(2) Mittie Quadrangle, Louisiana, U.S. Geological Survey, 1:24,000.
(3) Wolf Lake Quadrangle, Illinois, U.S. Geological Survey, 1:24,000.
(4) La Crescenta, Glendale and Sunland Quadrangles, California, U.S. Geological Survey, 1:24,000.
(5) Judson and Bryson Quadrangles, North Carolina, T.V.A., 1:24,000.
There is no significant difference between sample curves whose mean hysometric integrals differ by less than 8%. (From Strahler, 1952).

growth rate of an organ is a constant fraction of the specific growth rate of the whole organism, i.e.

$$Y = aX^b$$

where X = size of the organism
   Y = size of the organ
   a = a positive constant
   b = a constant

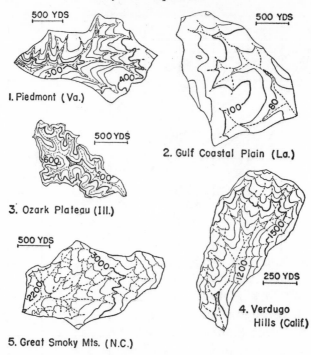

500 YDS
1. Piedmont (Va.)

500 YDS
2. Gulf Coastal Plain (La.)

500 YDS
3. Ozark Plateau (Ill.)

500 YDS
5. Great Smoky Mts. (N.C.)

250 YDS
4. Verdugo Hills (Calif.)

Fig. 6.24   Maps of representative drainage basins from the five areas summarized in Fig. 6.23. (From Strahler, 1952)

When two parts of such a system are related allometrically to the whole, they are related to each other by a power (i.e. logarithmic) function. Because Strahler stream order is logarithmically related to the process of stream discharge, it has been suggested that the laws of morphometry (involving the semi-logarithmic relationships between number of streams, mean stream length, mean stream area, mean stream slope, etc., on the one hand, and stream order, on the other) are really manifestations of allometric growth of stream systems. Some support for this idea comes from the fact that stream systems which are known to be actively expanding do so through the development of fingertip (first order) tributaries, which accords with the laws of morphometry. This process of growth is, however, complicated by the existence of a minimum threshold of catchment area which must be produced by excavation of the valley-side slope before a new first-order tributary can develop on it. Therefore, this development of first-order streams, and their ultimate evolution into second-order streams, and so on, implies that a hierarchy of basin orders develops, masked by a certain randomness, which tends to grow allometrically in a series of jumps to produce a continuous distribution with a number of peaks, each relating to one growth stage.

The third group of balancing-type equilibria in natural systems is that which is presumed to exist between opposing tendencies. This concept is deeply engrained in the classic concept of grade which embodies balances between erosion and deposition over given periods of time and between such nebulous quantities as "the capacity of the river to do work, and ... the quantity of work that a river has to do". A more secure example is of input being balanced by output in a basin hydrological cascading system wherein, when all the storages are filled (especially soil moisture storage), an input of precipitation (minus evaporation) will be equalled by the basin stream discharge output after a characteristic lag period (see Fig. 3.35).

As we have seen, a stream system may be considered as a system created by inputs of water (i.e. potential energy) at different elevations which are converted into kinetic energy. Because mean velocity does not usually significantly increase downstream, virtually all this kinetic energy is dissipated by friction between the water molecules or with the bed and bank material, and the heat removed from the system. Seven variables are mainly involved in downstream changes of channel geometry—width (W), depth (D), velocity (V), slope (S), sediment load (C), sediment calibre and hydraulic roughness (n)†, and discharge (Q). Three main relationships (1–3) govern the preservation of downstream equilibrium:

Continuity of discharge, i.e.

$$Q = WDV \tag{1}$$

Manning's equation of velocity, i.e.

$$V \propto (D^{2/3} S^{1/2})/n \tag{2}$$

Sediment transport equation, i.e.

$$C \propto [(VD)^{1/2} S^{3/2}]/n^4 \tag{3}$$

It has been previously pointed out that the maintenance of downstream equilibrium can be affected by a great variety of mutual adjustments of the

† When this is calculated indirectly by substitution in the Manning formula it is termed n' (See p. 209).

above seven variables, but these adjustments are additionally constrained by the operation of two opposing tendencies (4 and 5), such that the most probable distribution of energy along a stream course is a compromise between them. These constraints are:

(*a*) A constant (i.e. uniformly-distributed) rate of energy expenditure per unit area of stream bed, such that (for a unit weight of water);

S = channel slope (i.e. vertical fall of water in a unit distance)
VS = rate of fall
WD = weight of water per unit length of channel
$WDVS$ = rate of expenditure of energy per unit channel length = QS
QS/W = rate of expenditure of energy per unit area of channel bed.

$$\therefore QS/W = \text{constant} \tag{4}$$

Empirically, it is known that $W \propto Q^{1/2}$
$\therefore S \propto Q^{-1/2}$ (i.e. $z = -0.5$: See p. 209)

(*b*) Minimum total work expended in the system (i.e. a constant rate of energy expenditure per unit length of channel).

$$\text{i.e. } QS = \text{constant} \tag{5}$$

$\therefore S \propto Q^{-1}$ (i.e. $z \simeq -.999$)

A uniform distribution of energy expenditure could be met by having D and V uniform along the river, but because Q increases and debris is added by tributaries this simple arrangement is impossible. Therefore actual stream curves are intermediate between the slight concavity implied by (4) and the great concavity demanded by (5). The empirical average field relationship between slope and discharge is $S \propto Q^{-0.75}$ (i.e. intermediate between (4) and (5)) (Fig. 6.25) such that there is a gradual increase of W downstream, a concurrent smaller increase of D (so that W/D increases downstream), a slight increase of V and a decrease of hydraulic resistance (because decrease of particle size just more than offsets increases of resistance due to the increasing frequency of pools, riffles, bars and sinuosity downstream). There are no unique solutions by which the other variables accommodate the increase of discharge downstream, so that the above tendencies (operating through the mechanisms of scour and fill) are statistical ones and there is a spectrum of possible longitudinal profile forms, with a maximum probability of medium concavity.

The development of hierarchies of subsystems within open systems is sometimes another result of the interaction of opposing tendencies operating to produce subsystems of differing magnitudes and complexities in order to

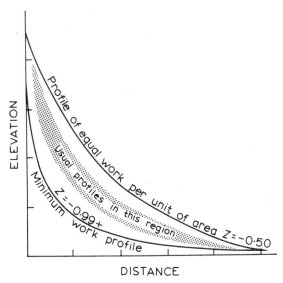

Fig. 6.25   Schematic longitudinal stream profiles, for a given total relief, depicting the zone of usual stream profiles compared with the theoretical minimum work and equal work profiles. (From Langbein and Leopold, 1964)

perform different (and often incompatible) tasks, and is reinforced by the existence of thresholds. A simple example of this is provided by the geometry of stream channels and slopes within an erosional drainage basin. Water falling on a land surface is most efficiently removed if it has to flow in an unconcentrated manner for only a short distance down a valley-side slope before reaching a channel, in which its removal is much more rapid and economical. From the point of view of unconcentrated surface runoff, therefore, it would be most efficient for the whole drainage area to be composed of infinitesimally short slopes each leading down to a basal stream channel (Fig. 6.26*a*). This would necessitate an immensely complex bifurcating network of small stream channels, however, and, when one considers the efficiency of channel flow, larger channels of greater width and depth remove water more efficiently per unit length than smaller ones. From this standpoint the greatest channel efficiency would be satisfied by the existence of one large channel draining the area (Fig. 6.26*b*). Obviously these conditions are incompatible and, in reality, represent the two extremes of a drainage hierarchy in which the number of smaller fingertip (i.e. first order) streams necessary for the economical collection of unconcentrated surface runoff is balanced by their progressive union to form larger (i.e. higher order) streams which remove the channel flow more efficiently. Thus a compromise between the two opposing tendencies is achieved by a hierarchy of stream orders (Fig. 6.26c). The lower limit to the hierarchy is imposed by a threshold which controls the lower size of first order fingertip tributaries. This must operate because it is not until a certain amount of surface runoff is gathered together that it becomes deep enough to be able to erode distinct channels. The length of first order channels is controlled by this limiting catchment area of first order basins. Of course, the above scheme is complicated in reality, firstly, by the fact that not all water flows off the surface of most drainage basins but some of it infiltrates, reaching the channels by underground flow; and, secondly, by the existence of eroded debris which influences

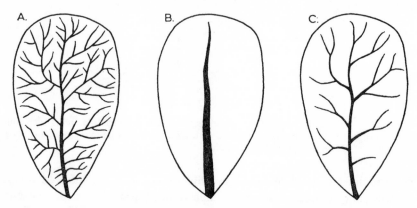

Fig. 6.26   Schematic illustration of the theoretical response of basin drainage density to different work efficiencies.

(a) A high drainage density required to allow the most economical unconcentrated surface runoff.
(b) A drainage pattern allowing the most efficient channel runoff.
(c) A usual type of drainage pattern representing a compromise between the two conflicting tendencies of (a) and (b).

both the character of unconcentrated surface runoff and the geometry of stream channels (and hence their efficiency for the removal of concentrated flow). Incidentally, this removal of debris is an irreversible process which introduces the kind of time-directed progressive change in a system which will be treated in the next chapter.

6.7 Thresholds

It is clear that the existence of thresholds separating different economies of system operation presents problems when one is considering the maintenance of an equilibrium state through the operation of negative feedback, because the passage of a system condition across a threshold is commonly irreversible (i.e. non-recoverable). Thus a small change in one critical variable may well force the system to adjust itself as a whole to a radically different dynamic equilibrium (see Fig. 6.1*h*). In chemistry these different economies are termed *phases*, and the equilibrium relationships are governed by the *Gibbs Phase Rule*.

It is interesting that the science of *statistical mechanics*, the first systematic account of which was given by Gibbs in 1902, is increasingly providing models (albeit mainly qualitative ones) within which many phenomena of interest to the physical geographer are being redefined. Statistical mechanics employs differential equations in the formulation of statistical laws regarding the behaviour of average system states, usually represented by large assemblages of small particles. According to this view of complex reality, the state of a system can be represented by a single image point in a complex phase space and expressed mathematically as a *phase function*—an equation in which the average system state is related to the variables forming the axes of the phase space. In figure 6.27 Y_1 is such a system state in a 3-phase space, expressed as a function of the three variables X_1, X_2, X_3. A system state is subject to a wide range of possible perturbations which involve changes in the phase function, and the controls over these trajectories of the image point (e.g. $Y_1 \rightarrow Y_1'$) are of prime interest to the student of the behaviour of complex systems. Such systems (i.e. being defined by many independent variables) possess a large number of degrees of freedom and it is consequently difficult to predict their trajectories. The situation may be simplified, however, if one can show that certain regions of the phase space are barred to possible trajectories (i.e. mature drainage basin geometrical system states cannot assume relationships involving angles of ground slope greater than the angle of repose of the debris—this limit is represented by the value X_3' in figure 6.27), or, in particular, if the image point tends to move on a restricted surface, or *manifold*, in the phase space defined on the assumption that certain independent variables are essentially constant in operation in some zones of the phase space (i.e. point Y_2 in figure 6.27) or that all variables change with some predictable regularity (i.e. such as to maintain a constant system energy level). Thus the two central problems of statistical mechanics are:

1 To determine under what conditions average time changes can be replaced by average phase-space changes, such that a phase-space trajectory may be viewed as a time trajectory. This is the *ergodic problem*, but the term is sometimes applied (as in Chapter 7) to express qualitatively the problem of using spatial variations of a system state to infer temporal changes.
2 To identify the form, or topology, of the manifolds.

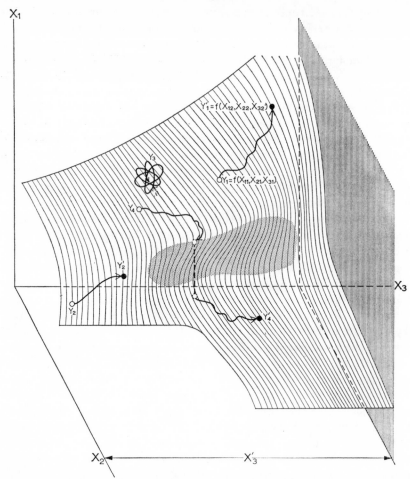

Fig. 6.27   A manifold in a 3-phase space containing system state trajectories illustrating
a simple trajectory ($Y_1 \rightarrow Y_1'$), a trajectory involving changes in only two variables ($Y_2 \rightarrow Y_2'$), homeostasis (Y_3), and thresholds ($Y_4 \rightarrow Y_4'$; and X_3'). A fuller description is given
in the text.

A pioneer example of the application of these principles to geomorphology
by Melton is treated in Chapter 7, in which a trajectory of changes in stream
frequency and drainage density is used to suggest a general model for the
development of mature drainage systems through time, under assumptions
of the stability of certain other basin variables. Other analogies have been
made between the results predicted from the processes associated with the
behaviour of large groups of particles, on the one hand, and the numbers of
stream segments in a large drainage network and the incremental changes of

direction involved in meandering, on the other. Thus statistical mechanics embodies a wide range of concepts relating to complex system states including equilibrium (point Y_3 in figure 6.27 illustrates homeostasis), time trajectories and thresholds. The latter are particularly interesting in that they may be variously viewed, for example, as discontinuities in the phase space separating it into zones of differing system economies or as deformations in the manifold. These can lead to catastrophic transitions ('flips') of the system state from one domain of operation into another, perhaps irreversible ones. Trajectory $Y_4 \rightarrow Y_4'$ in figure 6.27 illustrates such a flip in system state. It is easy to see why biologists interested in evolutionary processes are concerning themselves

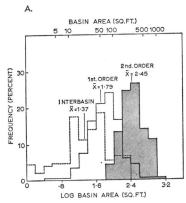
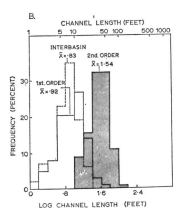

Fig. 6.28   Frequencies of (*a*) basin areas and (*b*) channel lengths of first and second order (plus interbasin areas and lengths) for the Perth Amboy badlands, New Jersey. (After Schumm, 1956)

with trajectories of system states, and particularly with the significance of catastrophic zones in the phase space, the passage of an organic system through which translates it from an economy exhibiting one set of relatively stable variations to another.

   An example of the possible operation of a threshold effect is that involving the permanent destruction of a vegetation cover by an infrequent runoff event of high magnitude, which ushers in a new runoff regime and texture of landscape dissection. The significantly different sizes of drainage basins of different order (Fig. 6.28) also suggest that when an expanding basin of a given order (*u*) exceeds a limiting size there is a jump in drainage network organization to produce a network of order $u + 1$. Even more subtle and far-reaching threshold effects have been suggested for landscape systems, for example that the tendencies for granites to form either mountains or plains in the tropics is possibly governed by a threshold value in the balance between the rate of surface erosion and the rate of subsurface chemical weathering.

The concept of thresholds has recently become of increasing significance in studies of valley-side slope geometry, and the threshold value of X_c has already been referred to (see Chapter 6.6). It has been observed, for example, that slips only occur in the San Gabriel Mountains of Southern California at slope angles of 38° or greater and in the London Clay at 10° or greater, implying some measure of stability below these threshold values. Debris slopes steeper than 28° in the southern Arizona desert behave in an unstable manner, and 26° seems to be the upper critical angle of stability (in terms of shallow sliding) for humid slopes covered with weathered debris comprising a mixture of rock rubble and coarse soil particles. Measurements of the maximum angles of straight slope segments in the Laramie Mountains of Wyoming reveal an upper limit of 33°, which is near the angle of repose for the rocky talus with less than 25% soil cover, and a lower limit of 18°, which is believed to be the threshold angle for debris in its most weathered and comminuted condition. The modal concentration of angles at 25°–28° is held to reflect the maximum angle of repose of completely saturated rock rubble and soil debris in an intermediate stage of breakdown.

6.8 Relaxation Paths

The study of relaxation paths of natural systems is very much in its infancy and very little specific information is yet available regarding the time patterns followed by the structure and behaviour of systems or system components when a change of input is causing a passage from one equilibrium condition to another. Theoretically such paths may show great variety in terms of such properties as 'false starts', 'reaction time', 'relaxation time', 'relaxation rate', 'overshooting', and the like. Some of the properties are illustrated in Fig. 6.29 where reaction time functions $[Y_x(t)]$ are plotted in respect of the application of different energy stimuli $[X_0(=0); X_1;$ etc.] to a recoverable system with zero *reaction time* (i.e. the time period (t_r) separating the change of input and the *beginning* of change in the system), a constant reaction time function, and zero relaxation time (i.e. the time period (t_x) between the beginning of system displacement and its 'settling down' to a new equilibrium state) (Fig. 6.29a); to a recoverable system with zero relaxation time, but with a variable reaction time function and with t_r dependent upon the sequence of inputs (i.e. the amount of system change and the reaction time are not simply dependent upon the nature of the isolated change of input) (Fig. 6.29b); and to a non-recoverable system where neither reaction nor relaxation times are equal to zero (Fig. 6.29c). Fig. 6.30 gives a simple illustration of the effect of relaxation time in damping the response of a system to a series of inputs to which it is continually attempting to adjust. Fig. 6.30a shows a number of exponential response curves exhibiting different rates of relaxation

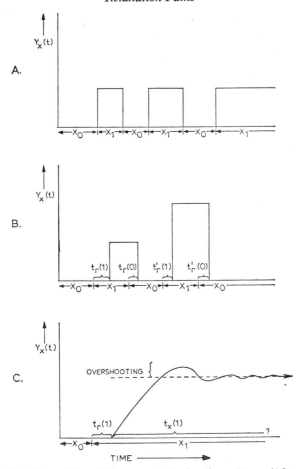

Fig. 6.29   Three different types of relaxation patterns. (After Klir and Valach, 1967). A full description is given in the text.

towards an equilibrium state after disturbance, and Fig. 6.30*b* compares the behaviour of a system with instantaneous response (0) (i.e. zero relaxation time) with that of another system having a medium response level (3) to the same series of random stimuli. This damping is due to a serial autocorrelation effect and the results are thus rather similar to those produced by filtering with moving averages (see Chapter 7.3); namely, an increase in the period of oscillation, a decrease in amplitude, a greater regularity of the distribution, and displacement of peaks. It is clear that the operation of relaxation time to produce response lags may, for example, so diminish the effects of individual inputs on the system as to reduce its operational level to below some important threshold.

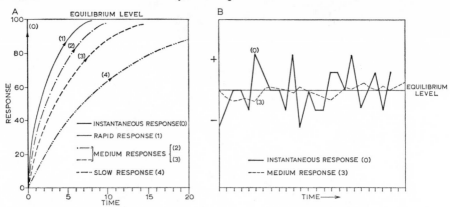

Fig. 6.30   The effect of relaxation time in damping the response of a system to a series of inputs (After Carlsson, 1968).
(*a*) A number of exponential response curves exhibiting differing relaxation times.
(*b*) A comparison of the behaviour of a system with instantaneous response (O) with that of another system having a medium response level to the same series of random stimuli.

Relaxation time in a system as a whole depends largely on four factors:

1 The resistance to change offered by the individual components of the system: i.e. their reaction and relaxation times. In geomorphology this is best illustrated by differences in resistance to erosion presented by various parts of a valley system (i.e. the valley-side slope, as distinct from the alluvial channel) or by various geological outcrops. The influence of system composition and linkages on relaxation paths can be represented graphically in terms of the deformation of materials by the *rheological models* shown in Fig. 6.31. Rheological models are mechanical analogues for the behaviour of different states of materials as they deform (e.g. are subject to *strain*) in passing from one equilibrium state to another, following the application of a new *stress*. Thus the basic elastic, viscous and crude (imperfect) plastic behaviour under stress are simulated, respectively, by that of a spring (coefficient of elasticity $= \alpha$), a permeable piston moving in a cylinder filled with fluid (viscosity $= \mu$) and a weight lying on a rough surface (coefficient of friction $= f$). Each type of material possesses its own relaxation pattern in response to the application of stress (in this case $= K$) over a given time period (t_0 to t_1). By comparing the strain patterns of real materials placed under experimental stress with those of the models, it is possible to infer the character and behaviour of such materials within different stress ranges. Most materials, like most systems, have such complex internal structures that their relaxation behaviour can only be

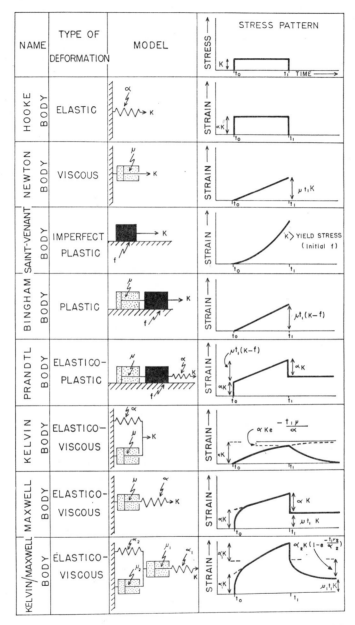

Fig. 6.31   The use of rheological models as analogues producing the different types of strain (i.e. relaxation) patterns exhibited by a wide range of solids.

interpreted by comparison with combinations of the more simple rheological models, for example, those simulating true plastic, elastico-plastic and elastico-viscous behaviour (Fig. 6.31).

2 The complexity of the system, involving the number of components (i.e. the phase space) and the frequency and nature of their linkages. Figure 6.31 illustrates the importance of linkages in a mechanical way, but a more interesting aspect of the influence of linkages has been suggested with respect to the equilibrium relationships among the major hydraulic variables in a downstream direction. It may well be that the very complexity and multiplicity of connections between these variables allows so many alternative possible adjustments between them, that equilibrium to changing load/discharge inputs can be achieved under a wide variety of conditions and thus is arrived at very quickly.

3 The magnitude and direction of the input = i.e. does the input change reinforce or counteract an existing system tendency?

4 The energy environment of the input. This topic has been covered extensively in Chapter 5, but it should be reiterated here that rapid input fluctuations tend to be 'filtered out' by systems with even short relaxation times, to produce simpler internal tendencies (see Fig. 6.30*b*).

The rate of relaxation can be more specifically understood with reference to:

1 The 'distance' of the parameter or system condition from its theoretical equilibrium state. One has seen, for example, how theoretical models are based on the simple assumption that rate of vertical denudation is directly proportional to distance above baselevel. Fig. 6.32 exemplifies this as the result of empirical experiment by showing the total net change in a model equilibrium beach (developed with a wave steepness (H/L) of 0·017) when the wave steepness was increased to 0·041. In a chemical reaction this concept is expressed by its *degree of advancement* (ξ), and the driving force (or 'potential') of the reaction in its approach to equilibrium is a derivative of the free energy (G) with respect to this degree of advancement (i.e. Reaction potential = $dG/d\xi$).

2 The ratio of the change-producing and the change-resisting forces. It has been shown, for example, that measured crustal uplift rates are some 8 times the average maximum denudation rates, making it unlikely that there can be the widespread development of landforms exhibiting the kind of equilibrium between 'endogenetic' and 'exogenetic' process suggested by W. Penck.

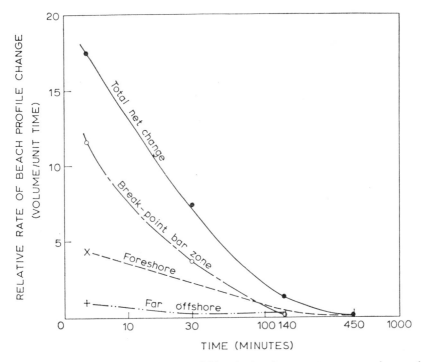

Fig. 6.32   Rates of relaxation on a model beach after the wave steepness was increased from 0·017 to 0·041. (After Scott. From King, 1959)

Obviously the patterns of relaxation paths can be viewed with respect to their 'sensitivity' and it has been suggested that the combination of high coefficients of correlation (r) and low standard errors (σ_{ys}) (see Fig. 2.1) is indicative of such sensitivity of the system variables to external changes. The problem of sensitivity is further complicated when the processes (i.e. energy inputs) are periodic with reversals, as well as variable. Measurements of net deposition (Y) were made daily for one low water condition at Virginia Beach, Virginia, during 25 days in June and July 1963. A multiple regression was used to 'explain' these depositional variations with respect to 14 independent variables measured every 4 hours during the period, such that the effect of each could be lagged with respect to 6 previous time periods (1. Falling to low water; 2. Rising and falling about high water; 3. Rising tide; 4. Falling to low water; 5. Rising and falling about high water; and 6. Rising tide) (Table 6.1). The results of this analysis show that the 6 most influential variables in determining net deposition on the lower foreshore are the slope angle of the lower foreshore ($X1$); the wave period ($X2$); the wave height ($X4$); the onshore wind velocity ($X6$); the angle of wave approach ($X9$); and the

TABLE 6.1

The Five Strongest Per Cent Reductions in Net Deposition at Lower Foreshore Stations. Sum of Squares Attributable to Combinations of Six Independent Variables for Each of Six Lag Periods. (From Harrison and Krumbein, 1964.)

Independent Variable Combinations												*Per cent Reduction in SS*
1	2	4	6	7	8	9	10	11	12	13	14	
Lag 1 1	2		6			9	10			13		73·31
(Falling 1	2		6			9		11		13		72·85
to Low 1	2		6			9				13	14	72·77
Water) 1	2	4	6			9				13		72·61
1	2	4	6			9	10					72·23
Lag 2 1			6				10	11	12		14	69·40
(Rising 1			6			9	10	11			14	68·04
and 1			6					11	12	13	14	67·76
Falling 1			6			9		11	12		14	67·62
About High 1		4	6					11	12		14	66·95
Water)												
Lag 3 1	2	4				9		11	12			72·41
(Rising 1	2	4				9		11			14	70·96
Tide) 1	2	4			8	9		11				70·78
1	2	4						11	12		14	70·56
1	2	4	6					11	12			70·45
Lag 4 1	2		6					11		13	14	61·26
(Falling 1			6			9		11		13	14	61·02
to Low 1	2	4						11		13	14	60·80
Water) 1			6				10	11		13	14	60·39
1	2	4			8			11			14	59·46
Lag 5 1	2	4	6			9	10					68·98
(Rising 1	2	4	6				10			13		68·80
and 1	2	4	6				10		12			68·75
Falling 1	2	4	6		8		10					68·56
about High 1	2	4			8			11		13		68·54
Water)												
Lag 6 1	2	4		7	8	9						76·14
(Rising 1	2	4	6		8	9						75·64
Tide) 1		4		7	8	9		11				72·36
1		4	6	7	8	9						72·34
1		4		7	8	9	10					72·06

Column legend (left to right):

Lower Foreshore Slope Angle · Wave Period · Wave Height · Onshore Wind Velocity · Offshore Wind Velocity · Parallel Wind Velocity · Wave Approach Angle · Longshore Current Velocity · Water Density · Rate of Water-level Rise · Rate of Water-level Fall · Water Table Depth

water density ($X11$); and that this combination of variables expresses itself most strongly during times of rising tide (Table 6.1). Other studies on the same beach showed that almost 50% of the variance observed in low-water slope of the foreshore could be explained by the following seven variables (in the listed order of importance, with the sign of r given): mean breaker power three hours previously (i.e. previous falling half tide) ($-$); the slope of the water table in the beach nine hours previously (i.e. previous rising half tide) ($+$); the slope of the beach water table at the low water ($+$); the ratio of mean sand diameter to the depth of water below wave trough at the low water ($-$); the mean angle of incidence of breaking waves to the shoreline nine hours previously ($+$); the slope of the beach water table three hours previously ($-$); and the total energy of the average breaker six hours previously (i.e. previous high tide). Besides indicating the importance of beach water table conditions in determining foreshore slope, this study highlights some of the special difficulties in analysing process-response systems with variable, periodic and reversed energy flows because of:

1 The short memory of the process variables which commonly change before beach equilibrium response has been achieved.
2 The replication among the process variables associated with feedback effects.
3 The relatively wide ranges in magnitude which the process variables may assume leading to a considerable variety of possible beach responses.

6.9 Equilibrium and Temporal Change

The problems associated with time-directed change have been present, implicitly or explicitly, throughout much of the foregoing discussion, and they form the topic for treatment in Chapter 7. It is true that much of what makes the world interesting to man derives from the dynamic, evolving or non-equilibrium aspects of his surroundings, but this interest is complicated by the difficulty of not always being able to distinguish between steady state changes and equilibrium relaxation on the one hand, and a *trajectory* of system states on the other. For example, some high intensity events are sufficient to transform the system to a new equilibrium state, so that change through time is sometimes merely a sequence of such transformations between different unstable equilibrium states. However, many such catastrophic changes are relatively rapidly effaced by a return to stable equilibrium. This results in a minimization of the landscape disorder resulting from many such changes over a long past history and the production of much of the observed order and regularity. In other words, the basic problem consists of recognising and isolating the progressive time trends of the dynamic and dynamic metastable equilibrium states (see Fig. 6.1).

Natural phenomena consist of a complex of interlocking systems having different threshold conditions, relaxation paths and time trajectories. When these are operated on by sequences of inputs having both random and time-controlled qualities, the result is a supersystem containing a hierarchical *plexus* of systems, subsystems and elements, each differently adjusted to the existing energy environment and, hence, each carrying an historical signal of different strength. The relaxation time of some natural systems is short (e.g. those involving hydraulic geometry and beach characteristics); for others it is longer (e.g. those involving drainage density); but for others it is long (e.g. changes in the relief of granite terrains induced by climatic change). Even one local landscape, for example, is made up of systems of widely differing behaviour, so that the landscape must be viewed as a *palimpsest* of systems, whose operation is interlocked, but whose history is superimposed. Just because some of the systems or their variables exhibit equilibrium, it does not necessarily imply that this is true for all others. Indeed one's whole attitude to natural systems stems from one's time perspective: where the relaxation time is short the timeless characteristics of the system seem dominant, where it is long the timebound or historical features preoccupy one.

REFERENCES

Abrahams, A. D. (1968), Distinguishing between the concepts of steady state and dynamic equilibrium in geomorphology; *Earth Science Journal*, **2**(2), 160–166.

Anderson, J. (1969), On general systems theory and the concept of entropy in urban geography; *London School of Economics, Graduate Geography Department, Discussion Paper 31*, 17.

Ashby, W. R. (1964) *An Introduction to Cybernetics* (University Paperback, Methuen, London), 295.

Blalock, H. M. and Blalock, A. B. (1959), Toward a classification of systems analysis in the social sciences; *Philosophy of Science*, **26**, 84–92.

Carlsson, G. (1968), Response inertia and cycles: A study in macrodynamics; *Acta Sociologica*, **11**, 125–143.

Carson, M. A. (1971), An application of the concept of threshold slopes to the Laramie Mountains, Wyoming: In Brunsden, D. (Ed.), *Slopes: Form and Process*, Institute of British Geographers, Special Publication No. 3, 31–48.

Carter, C. S. and Chorley, R. J. (1961), Early slope development in an expanding stream system; *Geological Magazine*, **98**, 117–130.

Cherry, C. (1966), *On Human Communication*, 2nd Edn. (M.I.T. Press, New York and London), 337.

Chorley, R. J. (1962), Geomorphology and general systems theory; *U.S. Geological Survey, Professional Paper 500-B*, 10.

Chorley, R. J. (1967), Models in Geomorphology; In Chorley, R. J. and Haggett, P. (Eds.). *Models in Geography* (Methuen, London), 57–96.

Clarke, D. L. (1968), *Analytical Archaeology* (Methuen, London), 684.

Connelly, D. S. (1968), *The Coding and Storage of Terrain Height Data: An introduction to numerical cartography* (M.Sc. Thesis, Cornell University), 140.

Cowan, T. A. (1963), On the very general character of equilibrium systems; *General Systems Yearbook*, **8**, 125–128.

Denbeigh, K. G. (1960–62), The science of continuously operating systems; *Inaugural Lectures, Imperial College, 1960–61 and 1961–62*, 1–18.

Dury, G. H. (1966), The concept of grade; In Dury, G. H. (Ed.) *Essays in Geomorphology* (Heinemann, London), 211–233.

Hack, J. T. (1960), Interpretation of erosional topography in humid temperate regions; *American Journal of Science*, **258-A**, 80–97.

Haggett, P. and Chorley, R. J. (1969), *Network Analysis in Geography* (Arnold, London), 348.

Hare, V. C. (1967), *Systems Analysis: A Diagnostic Approach* (Harcourt, Brace and World Inc., New York), 544.

Harrison, W. and Krumbein, W. C. (1964), Interactions of the beach-ocean-atmosphere system at Virginia Beach, Virgina; *U.S. Army Coastal Engineer Research Centre, Technical Memo No. 7*.

Harrison, W. (1970), Prediction of beach changes; *Progress in Geography*, **2**, 207–235.

Harvey, D. (1969), *Explanation in Geography* (Arnold, London), 521.

Holmes, C. D. (1964), Equilibrium in humid-climate physiographic processes; *American Journal of Science*, **262**, 436–445.

Horton, R. E. (1945), Erosional development of streams and their drainage basins: Hydrophysical approach to quantitative morphology; *Bulletin of the Geological Society of America*, **56**, 275–370.

Howard, A. D. (1965), Geomorphological systems—Equilibrium and dynamics; *American Journal of Science*, **263**, 302–312.

Khinchin, A. I. (1949), *Mathematical Foundations of Statistical Mechanics* (Dover, New York), 179.

King, C. A. M. (1959), *Beaches and Coasts* (Arnold, London), 403.

Kirkby, M. J. and Chorley, R. J. (1967), Throughflow, overland flow and erosion; *Bulletin of the International Association of Scientific Hydrology*, Year 12(3), 5–21.

Klir, J. and Valach, M. (1967), *Cybernetic Modelling* (Iliffe Books Ltd., London), 437.

Langbein, W. B. and Leopold, L. B. (1964), Quasi-equilibrium states in channel morphology; *American Journal of Science*, **262**, 782–794.

Leopold, L. B. and Langbein, W. B. (1962), The concept of entropy in landscape evolution; *U.S. Geological Survey, Professional Paper 500–A*, 20.

Leopold, L. B., Wolman, M. G. and Miller, J. P. (1964), *Fluvial Processes in Geomorphology* (Freeman, San Francisco), 522.

Lotka, A. J. (1956), *Elements of Mathematical Biology* (Dover, New York), 465.

Machol, R. E. (Ed.), (1965), *System Engineering Handbook* (McGraw-Hill, New York).

Mackin, J. H. (1948), Concept of the graded river; *Bulletin of the Geological Society of America*, **59**, 463–512.

Marujama, M. (1963), The second cybernetics: Deviation-amplifying mutual causal processes; *American Scientist*, **51**, 164–179.

Medvedkov, Y. (1969), Entropy: An assessment of potentialities in geography; *Commission on Quantitative Methods, International Geographical Union, London, August 1969*.

Melton, M. A. (1958), Geometric properties of mature drainage systems and their representation in an E_4 phase space; *Journal of Geology*, **66**, 35–54.

Melton, M. A. (1965), Debris-covered hillslopes of the Arizona desert; *Journal of Geology*, **73**, 715–729.

More, R. J. (1967), Hydrological models and geography; In Chorley, R. J. and Haggett, P. (Eds.), *Models in Geography* (Methuen, London), 145–185.

Rosen, R. (1967), *Optimality Principles in Biology* (Butterworths, London), 198.

Ruxton, B. P. (1968), Order and disorder in land; In Stewart, G. A. (Ed.), *Land Evaluation* (Macmillan, Melbourne), 29–39.

Scheidegger, A. E. and Langbein, W. B. (1966), Probability concepts in geomorphology; *U.S. Geological Survey, Professional Paper 500-C*, 14.

Schumm, S. A. (1956), The evolution of drainage systems and slopes in badlands at Perth Amboy, New Jersey; *Bulletin of the Geological Society of America*, **67**, 597–646.

Strahler, A. N. (1950), Equilibrium theory of erosional slopes, approached by frequency distribution analysis; *American Journal of Science*, **248**, 673–696 and 800–814.

Strahler, A. N. (1952), Hypsometric (area-altitude) analysis of erosional topography; *Bulletin of the Geological Society of America*, **63**, 1117–1142.

Towler, J. E. (1969), *A Comparative Analysis of Scree Systems developed on the Skiddaw Slates and Borrowdale Volcanic Series of the English Lake District* (Unpublished B.A. Dissertation, Cambridge University).

Waddington, C. H. (Ed.) (1968), *Towards a Theoretical Biology: 1 Prolegomena* (Edinburgh University Press), 234.

Waddington, C. H. (Ed.) (1969), *Towards a Theoretical Biology; 2 Sketches* (Edinburgh University Press), 351.

Wall, F. T. (1965), *Chemical Thermodynamics;* 2nd Edn. (Freeman, San Francisco), 451.

Woldenberg, M. J. (1968), *Hierarchical Systems: Cities, rivers, alpine glaciers, bovine livers and trees* (Ph.D. Dissertation, Columbia University), 150.

Wolman, M. G. and Miller, J. P. (1960), Magnitude and frequency of forces in geomorphic processes; *Journal of Geology*, **68**, 54–74.

Wolman, M. G. (1955), The natural channel of Brandywine Creek, Pennsylvania; *U.S. Geological Survey, Professional Paper 271*, 56.

Young, O. R. (1964), A survey of general systems theory; *General Systems Yearbook*, **9**, 61–80.

7: *Change in Systems*

Of systems, the minutest crumb
Must Be, Behave, and then Become.
This Principle the space traverses.
From Atoms up to Universes.
And systems that are not malarky
Must find their place in this hierarchy.

KENNETH BOULDING

7.1 Introduction

At the conclusion of the previous chapter we gave the impression that it is possible to distinguish clearly between time-directed trajectories of system states involving non-recoverable change or evolution, and the more ephemeral compensating changes associated with equilibrium conditions. Any recognition of the distinction between timeless reversible and timebound irreversible change, even if the latter is only a small component of the former, implies that in some respects time may be *anisotropic* and can be regarded as an arrow. This concept embodies a great deal of classical thought in natural philosophy and is opposed to the more recent notion that time is a mere co-ordinate, providing simply a framework within which events can occur. In short that 'time is not a process in time'. Ever since the age of Archimedes there has been a trend towards the elimination of time from natural philosophy and this has only recently been counteracted by the ideas embodied in the notion of statistical irreversibility, associated with the tendency for entropy to increase progressively in processes of the thermodynamic type.

Some scholars find the distinction between timeless and timebound changes impossible to sustain and regard 'being' and 'becoming' as aspects of an identical process, linked by 'behaving' (as the verse at the head of this chapter illustrates). According to this view systems have a constant architecture in time which can be represented by a spiral of cause and effect passing from 'becoming' (i.e. developing or evolving) at one integrative level of organization, to 'being' (i.e. adopting a characteristic structure or morphology) at a higher level, to 'behaving' (i.e. operating in a self-justifying, equilibrium manner) at a still higher level, and so on (Fig. 7.1). Thus a progressive integration of organization within the system leads inevitably to an irreversible evolution. Although this idea applies most obviously to the organic world, its general implications are interesting.

Another important notion, involving the dynamism of time, is that the character of reality alters according to the timescale in which it is viewed.

251

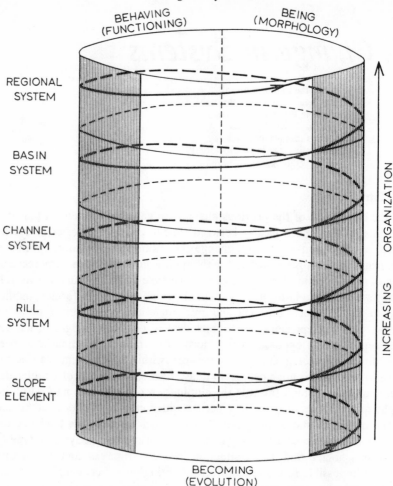

REGIONAL
SYSTEM

BASIN
SYSTEM

CHANNEL
SYSTEM

RILL
SYSTEM

SLOPE
ELEMENT

BEHAVING
(FUNCTIONING)

BEING
(MORPHOLOGY)

INCREASING   ORGANIZATION

BECOMING
(EVOLUTION)

Fig. 7.1   An illustration of the view that systems possess a constant architecture in time, represented by a spiral of cause and effect. (After Gerard, 1964)

Thus both the type of change which one is prepared to envisage and whether features of the systems are viewed as causes (i.e. independent variables) or effects (i.e. dependent variables) can be clearly shown to be functions of timespan. It has been suggested, for example, that the types of changes to which landform geometry (e.g. stream channel gradient) is susceptible can be best identified and emphasized by adopting such terms as cyclic (i.e. geological), graded (i.e. subjected to periodic fluctuations) and steady (i.e. unchanging) for the timespans which exemplify them (Fig. 7.2). The changing status of system variables is illustrated for drainage basins in Table 7.1

TABLE 7.1

Drainage basin variables	Status of variables during designated time spans		
	Cyclic	*Graded*	*Steady*
1 Time	Independent	Not relevant	Not relevant
2 Initial relief	Independent	Not relevant	Not relevant
3 Geology (lithology, structure)	Independent	Independent	Independent
4 Climate	Independent	Independent	Independent
5 Vegetation (type and density)	Dependent	Independent	Independent
6 Relief or volume of system above base level	Dependent	Independent	Independent
7 Hydrology (runoff and sediment yield per unit are a within system)	Dependent	Independent	Independent
8 Drainage network morphology	Dependent	Dependent	Independent
9 Hillslope morphology	Dependent	Dependent	Independent
10 Hydrology (discharge of water and sediment from system)	Dependent	Dependent	Dependent

(From Schumm and Lichty, 1965)

TABLE 7.2

River Variables	Status of variables during designated time spans		
	Geologic	*Modern*	*Present*
1 Time	Independent	Not relevant	Not relevant
2 Geology (lithology and structure)	Independent	Independent	Independent
3 Climate	Independent	Independent	Independent
4 Vegetation (type and density)	Dependent	Independent	Independent
5 Relief	Dependent	Independent	Independent
6 Paleohydrology (long-term discharge of water and sediment)	Dependent	Independent	Independent
7 Valley dimension (width, depth, and slope)	Dependent	Independent	Independent
8 Mean discharge of water and sediment	Indeterminate	Independent	Independent
9 Channel morphology (width, depth, slope, shape, and pattern)	Indeterminate	Dependent	Independent
10 Observed discharge of water and sediment	Indeterminate	Indeterminate	Dependent
11 Observed flow characteristics (depth, velocity, turbulence, *et cetera*)	Indeterminate	Indeterminate	Dependent

(From Schumm and Lichty, 1965)

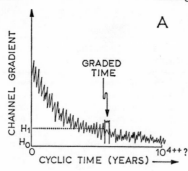

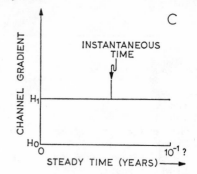

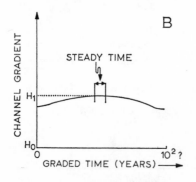

Fig. 7.2   Diagrams illustrating changes of stream channel gradient at a point through (*a*) cyclic, (*b*) graded and (*c*) steady time. (After Schumm and Lichty, 1965)

and for river variables in Table 7.2. Taking basin relief (6) as an example from Table 7.1, in a long-term (cyclic) sense it is a variable depending on variables 1 to 4, but in shorter timespans it is better treated as an independent and fixed parameter. Thus, as the timescale on which we view a system changes, so the variables in operation can be thought of as independent, dependent, or even irrelevant.

7.2 Causes of Non-Reversible Change within Systems

The nature of long-term system changes has formed much of the basis for classical investigations in the earth sciences, commonly having to do with questions of evolution. (It should, perhaps, be pointed out here that although this emphasis on evolutionary processes in the earth sciences derives directly from the work of Charles Darwin, the idea of evolution involved is an over-simple one. That is to say, evolution, particularly with respect to landscapes, is taken to mean any progressive change and is divorced from the ideas of increasing specialisation or diversification in form which were inherent in Darwin's thesis.) In the present context these changes may be considered to fall in four classes: firstly, time-directed changes of energy and its distribution within the system, such as those with which the Davis cycle of erosion was concerned; secondly, those due to significant changes either of input or of input/output relationships; thirdly, those associated with changes of sub-systems integration; and, fourthly, those resulting from the progressive influence of large storages with very lagged outputs.

An important class of time-directed changes of energy and mass distribution within systems is that associated with the deviation-amplifying effects of major positive feedback loops. As has been already pointed out, such major effects are for the most part lacking in physical systems, while forming a keystone of organic evolutionary ideas. It will be necessary to return to this theme in Chapter 8.

However, one important tendency for progressive change in isolated physical systems is that, given an initial amount of initial free, or potential, energy within the system, they develop toward states with maximum entropy. Entropy, as we have already pointed out, is an expression for the degree to which energy has become unable to perform work. An increase of entropy implies a trend toward minimum free energy and increasing inability to perform work. Hence, in an isolated system there is a tendency for the initial differences within the system to be removed as mass and energy cascade irreversibly from one subsystem to another. This process is expressed by the Second Law of Thermodynamics which, in its classic form, is formulated for isolated systems. In such cases, therefore, the change of entropy is always positive, associated with a decrease in the amount of free energy, or, to state this another way, with a tendency toward progressive destruction of existing order or differentiation. Thus, one can see that Davis' view of landscape development contains certain elements of isolated system thinking—including, for example, the idea that a given amount of potential energy is provided by the initial uplift and that, as degradation proceeds, the energy of the system decreases until, at the stage of peneplanation, there is a minimum amount of free energy as a result of the levelling down of topographic differences. The Davisian peneplain, therefore, may be considered as logically equivalent to the condition of maximum entropy, since general energy properties are more or less uniformly distributed throughout the system, with potential energy approaching zero. The positive change of entropy, and the associated negative change of free energy, implies the irreversibility of events within isolated systems. This again bears striking similarities to the general operation of the Davisian geomorphic cycle. The belief in the sequential development of landforms in sympathy with the reduction of relief, involving the progressive and irreversible evolution of almost every facet of landscape geometry, including valley-side slopes and drainage systems, is in accord with isolated system thinking. Although 'complications of the geographical cycle' can, in a sense, put the clock back, nothing was considered by Davis as capable of *reversing* the clock. The putting back of the clock by uplift, therefore, came to be associated with a release, or an absorption into the new isolated system, of an increment of free energy, which would subsequently be progressively dissipated through degradation.

In closed or open systems, on the other hand, the increase of entropy is not dominant, but is simply a tendency which serves to dissipate energy inputs. With time, the mass of the landscape is reduced and there are, inevitably, progressive changes in at least some of its absolute geometrical properties. One cannot, however, assume (as Davis did) that *all* the geometric properties are necessarily involved in these sequential changes: the falseness of this proposition has been demonstrated by many of the examples given in Chapter 6. The position is basically this: there are constant adjustments made in every landscape to new steady-state conditions, but these are superimposed upon a general tendency for change which is associated with the reduction of average relief through time. Although relief declines in this way, it does not *necessarily* involve a sympathetic change in all the other features of landscape geometry. If we take the example of drainage density, then it is clear that this variable is controlled by a number of factors of which relief (whether relative or absolute) is only one. Recent work, in fact, seems to indicate that relief and allied considerations of average land slope have, in all probability, only a relatively slight influence over drainage density and this control may well be masked or altogether counteracted by other more immediately important factors (for example, rainfall intensity and surface resistance) which are not so obviously susceptible to changes with time. Davis' view of landscape evolution was that the passage of time, of necessity, imprinted recognizable, significant and progressive changes on every facet of landscape geometry. The recognition, however, that landscape forms represent a steady-state adjustment with respect to a multiplicity of controlling factors obliges one to take a less rigid view of the evolutionary aspects of geomorphology. When a geometrical form is controlled by a number of factors, any change of form with the passage of time is entirely dependent upon the net results of the effect of time upon those factors. Some factors are profoundly affected by the passage of time, others are not; some factors act directly (using the term in the mathematical sense) upon the form, others inversely; some factors exercise an important control over form aspects, others a less important one. Thus, if a particular geometrical feature of the landscape is primarily controlled by a factor the action of which does not change greatly with time, or if the changes in factors having direct and inverse controls tend to cancel out each other, then the resulting variation in geometry may itself be small—perhaps insignificant. In short, although the equilibrium configuration of terrain is necessarily subject to progressive change as the available relief is reduced, significant change implies evolution in the nature of a major geometric constraint on landscape processes.

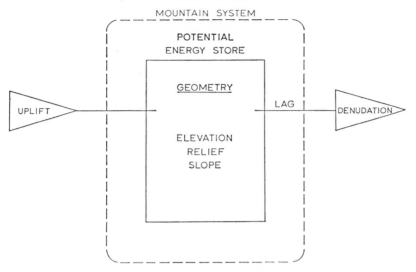

Fig. 7.3   Canonical representation of a mountain terrain system showing how land-scape geometry is related to the uplift and the (lagged) denudation.

Changes in such constraints are usually slow in comparison with changes in fluvial processes, so that, as Chapter 6 showed, forms of equilibrium are possible despite the evolution of the constraints, such that to a large extent present landforms are independent of previous evolution.

A second class of evolutionary changes to which physical systems are subject involves input/output changes resulting from progressive, long-continued changes of system inputs. The analysis of input changes is treated later in this chapter, but it is convenient here to draw attention to the controls which the relaxation times within the system exert over the ability of its various system components to reflect progressive changes of inputs. Where relaxation times are short, as with the atmospheric circulation, a progressive change of energy inputs will be more-or-less faithfully reflected in changes in the magnitude and operation of system components. In a terrain system, however, the ability of landform variables to adjust to changing energy throughputs depends upon the ability of denudational outputs (i.e. rates of degradation) to adjust to inputs of potential energy (i.e. rates of uplift) (Fig. 7.3). Many of the geomorphic concepts of W. Penck, for example, depended on his assumption that numerous classes of landforms develop under conditions of equilibrium between uplift and denudation. Fig. 7.4 shows a flow chart illustrating a hypothetical computer programme designed to simulate interactions between local erosion, deposition and isostatic

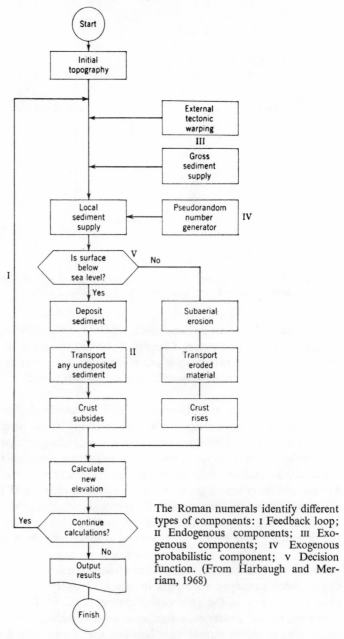

The Roman numerals identify different types of components: I Feedback loop; II Endogenous components; III Exogenous components; IV Exogenous probabilistic component; V Decision function. (From Harbaugh and Merriam, 1968)

Fig. 7.4   Simplified flow chart of a hypothetical computer programme designed to simulate the interactions between erosion, deposition and isostasy.

compensation. Such simplified models ignore, firstly, the fact that a given mean denudation rate can encompass a variety of morphological transformations, and, secondly, that the adjustment of denudation rates to uplift rates involves a lag as the effects of increased denudation rates spread through the drainage systems and to the highest divides. It has been estimated from sediment discharge rates from drainage basins of about 1500 square miles in area that mean maximum denudation rates are about 3 feet per 1000 years, whereas present maximum measured rates of orogenic uplift average some 25 feet per 1000 years. It has been suggested recently that this large disparity between denudation and uplift implies that either a rate of uplift must remain constant for a very long time period (e.g. tens of millions of years), or that local relief must achieve some 20,000 feet, before rates of denudation on the summits are equalized and a steady state occurs in landform development. Consequently the idea that widespread balances can occur between uplift and denudation yielding equilibrium relief, slope and elevation forms appears unlikely.

A third source of progressive change in systems concerns the history of the linkage of variables which are associated with the mutual operation of subsystems. One aspect of this was expressed in the verse which opened this chapter suggesting that, especially in organic systems, the 'being' of a system automatically implies a given 'behaving' which, in turn, automatically implies a given evolutionary 'becoming' leading to a higher level of integrated 'being', and so on. A more common integrative change in non-organic systems occurs when a decline in energy throughput causes operational linkages between system parameters to fall below minimum threshold levels. This leads to a measure of dynamic isolation between the cascade subsystems. When this occurs the structure of the system begins to disintegrate as groups of morphological parameters increasingly develop more-or-less independently of mutual regulation, so that there is a progressive destruction through time of the stable correlation links which were developed during a more energetic phase of equilibrium operation of the system. Two examples of such a time-bound disintegration have already been given. Fig. 6.7 showed how the removal of a basal stream, as a result of meandering, a general decrease in discharge, or valley-side retreat resulting from the progressive removal of basin mass, breaks the links between group of slope and stream parameters. Similarly, Fig. 6.8 suggested that the decrease of frost weathering on cliff faces in the Lake District composed of resistant Borrowdale Volcanics accompanying the replacement of periglacial conditions by a more temperate climate, cut off the supply of falling debris feeding the scree slope and led to more and more disorganized morphological relationships on this formation. This contrasts with the continuing high level of correlation structure exhibited by the less-resistant Skiddaw Slates which are still highly susceptible to weathering under modern climatic conditions. It is difficult both to predict the course of such a break up in detail and to infer the subsequent development of essentially isolated parts of the system, because the sequence of events depends not only on the sign of the correlation linkages in the system but also upon the relative strength of each link in the feedback loops.

A final type of progressive change to which a system may be subject results when storages are large and the lag of removal of mass or energy from storage is great. It has been suggested, for example, that long-term climatic change could result from random transfers of heat energy from the atmosphere to the large oceanic storage from which it is released only slowly, as a result of which the whole thermal economy of the globe may change. It is clear that this mechanism of change is similar to that of mass storage through orogenic uplift and its relatively slow dissipation by erosion which was treated earlier in this chapter.

7.3 Analysis of Observed Trends

The most important mechanism which causes process/response systems to change through time is change of input. The analysis of time trends exhibited by complex sequences involves the identification and removal of major long-term trends, commonly by curve-fitting or smoothing techniques, leaving the residual or *filtered* variations forming a *stationary time series* to be further analysed in terms of 'cyclical movements' (i.e. those having a roughly oscillatory change), 'periodic movements' (i.e. those having regular sequences of repetitions) and the remaining irregular random 'noise'. These time components are clearly illustrated by Fig. 7.5 in which a linear trend is added to a broad cyclic tendency, a seasonal periodicity and an irregular noise level to produce the composite graph. Viewed from the reverse, this process is one of *time filtering* whereby the filtering out of T leaves a residual stationary time series 'signal' of $C + S + I$, from which, in turn, the filtering out of C leaves $S + I$, and so on. It is usually most convenient to employ a best-fit linear (i.e. first-order polynomial) regression to identify a long-term trend in a series of time inputs, the significance of trend of which can be estimated by the use of the statistic 't' in testing the null hypothesis that the regression coefficient is not significantly different from zero. However, the data can be increasingly well approximated by the fitting of successively high order functions and the 20th order polynomial in Fig. 7.6 provides a significantly better fit to the data than does the first order regression. On the other hand, the more complex the fitted regression obviously the less clearly it identifies the long-term trend. Such time trends are exhibited by the channel aggradation in certain California rivers where, during the period 1849–1913,

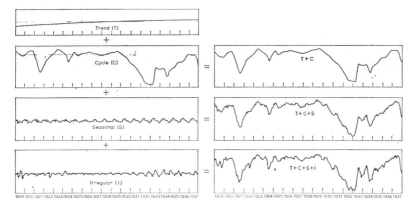

Fig. 7.5   Graph of pig iron production in the United States (1919–1937) (T + C + S + I), broken down into its major trend (T), as well as cyclic (C), seasonal (S) and irregular (I) components, together with combinations of these. (After Croxton and Cowden, 1948)

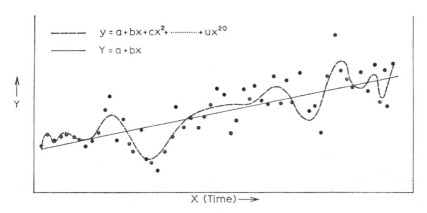

Fig. 7.6   A time series of points to which have been fitted a first-order (i.e. linear) and a 20th-order polynomial regression. (After Agterberg, 1967)

some 62% of the fill was provided by hydraulic mining operations (Fig. 7.7*a*) and in the Rio Grande between 1895 and 1935 due to vegetation changes and gullying (Fig. 7.7*b*). During the period 1880 to about 1900 there was a period of excessive erosion and arroyo cutting in the American South-West and a dissection of the rainfall inputs for this period is illustrative of the complexity of input analysis. Fig. 7.8 shows that, although there was no noticeable trend in annual rainfall, a distinct increase of lower intensity rains (0·01–0·49 in/day) occurred during this period, being more marked

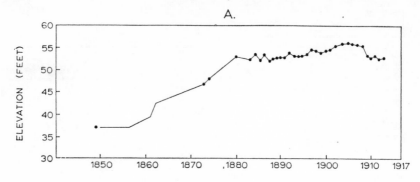

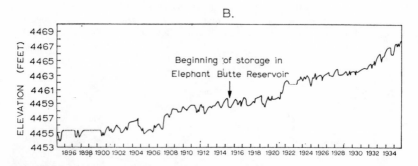

Fig. 7.7   Plots of fluctuations in elevation of low-water surfaces for:

(*a*) Yuba River at Marysville, California, due to deposition of mining debris (Data from Gilbert).

(*b*) Rio Grande River at San Marcial, some 40 miles upstream of Elephant Butte Dam.

(From *Fluvial Processes in Geomorphology* by Luna B. Leopold, Gordon Wolman and John P. Miller. W. H. Freeman & Company. Copyright © 1964)

during the summer months. Although it is not directly possible to associate this aspect of input change with erosional outputs after 1880 (increased grazing by cattle probably being of paramount importance) the relatively low occurrences of grass-producing summer rains during this period of heavy cattle stocking probably facilitated accelerated surface erosion.

Another method of eliminating small-scale oscillations of input to reveal long-term trends or irregular cycles involves the use of smoothing functions. The most common of these is the moving mean which operates as a *low pass filter*, retaining changes of long wavelength while filtering out the short ones. The fineness of the filter 'mesh' is determined by the number of means used and Table 7.3*a* illustrates the use of a 5-year moving mean smoothing

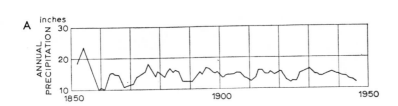

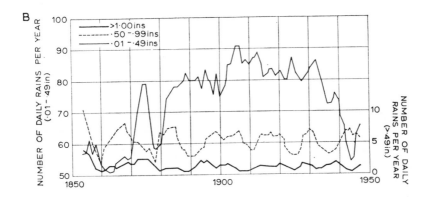

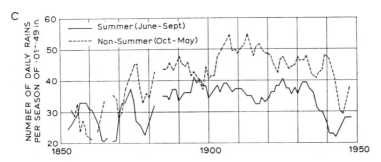

Fig. 7.8   Rainfall at Santa Fe, New Mexico, plotted as five-year moving means. (After Leopold, 1951)

(*a*) Annual precipitation.
(*b*) Number of daily rains per year for each of three magnitude classes.
(*c*) Number of daily rains per season of ·01–·49 inches.

TABLE 7.3

(A) 5-YEAR MOVING MEAN				(B) 5-YEAR NORMAL SMOOTHING FUNCTION			
	SMOOTHING FUNCTION	TIME SERIES	SMOOTHED		SMOOTHING FUNCTION	TIME SERIES	SMOOTHED
	·2	$P_1 = 28$			·03	$P_1 = 28$	
	·2	$P_2 = 23$			·23	$P_2 = 23$	
Central	·2	$P_3 = 21$	23·6*	Principal →	·48	$P_3 = 21$	22·39
Value →	·2	$P_4 = 24$	22·0	Weight	·23	$P_4 = 24$	22·70
	·2	$P_5 = 22$	—		·03	$P_5 = 22$	—
		$P_6 = 20$	—			$P_6 = 20$	—
		$P_7 = 17$	—			$P_7 = 17$	—
		$P_8 = 18$	—			$P_8 = 18$	—
		$P_9 = 29$				$P_9 = 29$	
		$P_{10} = 35$				$P_{10} = 35$	

$* = (28 \times ·2) + (23 \times ·2) + (21 \times ·2) + (24 \times ·2) + (22 \times ·2)$

or $\dfrac{28 + 23 + 21 + 24 + 22}{5}$

function about a central value. Fig. 7.9 gives 30-year moving means for some January temperature trends since 1800 and Fig. 7.10 the January temperatures for New York City (1871–1958) smoothed by a 10-year moving mean, in which a strong 10–11 year cycle appears to be superimposed upon a general warming trend. A comparison of the effects of moving means of different length is given in Fig. 7.11 with respect to annual precipitation at Omaha, Nebraska (1871–1940). The use of moving mean filtering has certain disadvantages, however, as a method of generalizing system inputs in that:

1 There is a loss of marginal data, which increases with the length of the filter.

2 'Reversals of polarity' (i.e. highs changed to lows, and vice versa) (Fig. 7.12, point A), and 'phase shifts' (i.e. displacement of peaks) of the shorter period fluctuations can occur.

3 A moving mean can sometimes artificially generate a false periodicity. Fig. 7.13 illustrates how a periodicity produced by the application of a 10-year moving mean (B) to annual temperatures for central England (1698–1952) (A) also occurs when a similar moving mean is applied (D) to the same data rearranged in an entirely random order (C).

One way of mitigating some of these difficulties is to apply a smoothing function of various weights decreasing outward from a 'principal weight'. Table 7.3b shows the application of a 5-year 'normal curve' smoothing function based on values proportional to the ordinates of the normal probability

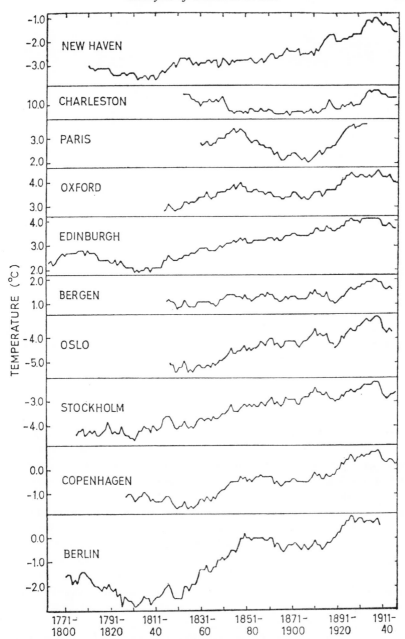

Fig. 7.9 January temperature trends since 1800 (30-year moving means) for ten stations in Europe and North America. (After Lysgaard. From Barry and Chorley, 1971)

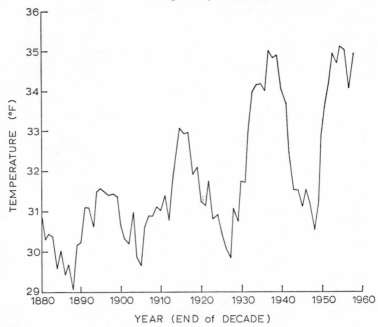

Fig. 7.10   January temperature trends from 1871 to 1958 (10-year moving means) for New York City. (After Spar. From Dorf, 1960)

curve. Fig. 7.12 compares the application of a normal curve smoothing function with moving mean and exponentially-weighted smoothing functions—the latter exhibiting marked phase shifts at B and C. Fig. 7.14 illustrates the effectiveness of a normal curve smoothing function in identifying both a general trend and an approximate 30-day cycle in barometric pressures at Washington, D.C. It should be pointed out that such smoothing functions can be applied to spatial smoothing, by weighting locations according to their distance from the central origin. Such generalization, used for example in smoothing meteorological pressure charts, bears an important relationship to the types of spatial filtering described in Chapter 4.

The methods described above are commonly employed to identify long-term and irregular cyclical trends in system components. However, many temporal tendencies are concerned with regular periodic changes and the analysis of these involves the use of harmonic analysis and its extensions. Fig. 7.15 shows the features of periodic (*a*) and random (*b*) oscillations, and suggests how the fitting of a periodic function to a composite stationary series (*c*) can separate a periodic tendency from more random fluctuations. In this way harmonic analysis is valuable both in providing mathematical

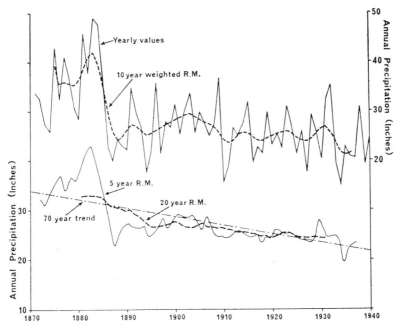

Fig. 7.11   Annual precipitation data from Omaha Nebraska (1871–1940), showing the relative smoothing effects of three types of moving means (or running means—RM), together with the general 70-year trend. (After Foster. From Barry, 1969)

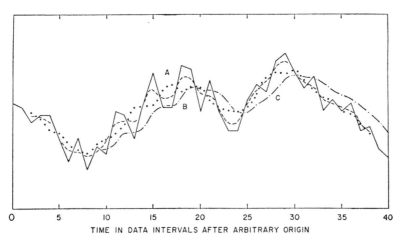

TIME IN DATA INTERVALS AFTER ARBITRARY ORIGIN

Fig. 7.12   An actual time series (solid line), and the same series smoothed by means of an equally-weighted moving mean (dotted line), by a normal-curve smoothing function (dashed line), and by exponential smoothing (dot-dashed line). Note the reversal of polarity (Point A on the equally-weighted moving mean) and the phase shifts (Points B and C on the exponential smoothing). (From Holloway, 1958)

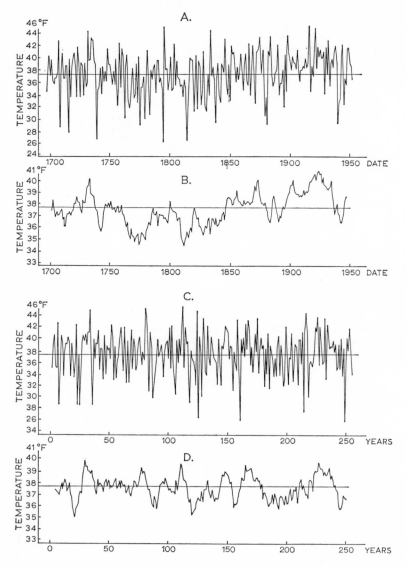

Fig. 7.13   The application of a 10-year moving mean to annual temperatures for central England, 1698–1952. (From Lewis, 1960)

(*a*) The plot of annual temperatures.
(*b*) Temperatures in (*a*) smoothed by the application of a 10-year moving mean.
(*c*) The temperatures in (*a*) rearranged in an entirely random manner.
(*d*) Temperatures in (*c*) smoothed by the application of a 10-year moving mean.

components which assist in more closely approximating the description of complex time trends possessing periodic components, and in the converse sense of being able to identify or 'filter out' the harmonic characteristics of the dominant periodicities which may be useful in suggesting the underlying mechanisms of change. An example of simple harmonic analysis, employing the cosine curve, is given in Fig. 7.16*a* where the form of the curve is defined by its amplitude (A: which in this case = unity) and the 'frequency' or

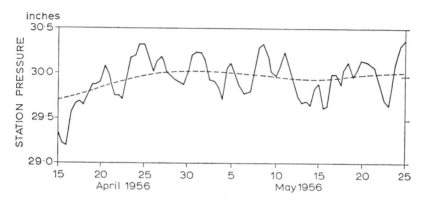

Fig. 7.14   The application of a low-pass filter (filter output dashed) to a time-distribution of atmospheric pressure (solid line) at the Washington National Airport, April–May 1956. (After Holloway, 1958)

number of waves (k) (in this case = 2) in the basic interval ($\theta_n = 2\pi$), the latter determining the wavelength ($= \theta_n/k = 2\pi/2 = \pi$). A further parameter used to define a periodic function is the 'phase shift' (Fig. 7.16*b*) which is the phase angle (Φ_k) divided by the frequency (k). This phase shift defines the distance of the first crest from the ordinate, which in Fig. 7.16*b* is $\dfrac{\pi}{6}\left(= \dfrac{\Phi_k}{k} = (\pi/3)/2\right)$. Thus the general equation for the curve in Fig. 7.16*b* is:

$$Y = A \cos (k\theta - \Phi_k),$$

or, more specifically,

$$Y = \cos [2\theta - (\pi/3)].$$

Much more complex curves can be approximated by a series of waves (commonly sine or cosine) of various amplitudes, frequencies and phases,

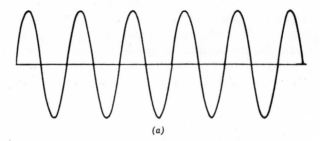

(*a*)

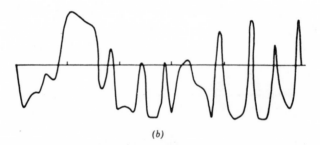

(*b*)

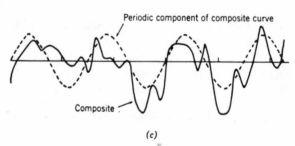

(*c*)

Fig. 7.15   Types of oscillatory behaviour. (After Preston and Henderson. From Harbaugh and Merriam, 1968)

(*a*) Periodic oscillation.
(*b*) Random oscillation devoid of periodic component.
(*c*) Composite oscillation, containing periodic and random fluctuations.

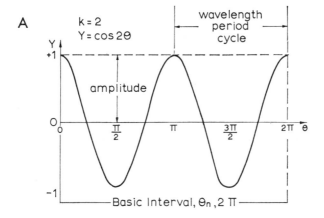

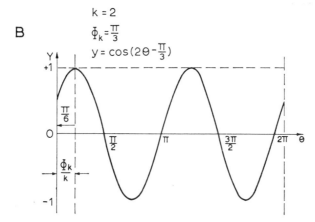

Fig. 7.16   Parameters involved in harmonic analysis. (After Rayner). A description is given in the text

and the Fourier theorem states that, no matter how complicated the fluctuations in the data, they can be accounted for by the superimposition of a number of simple component waves. Fig. 7.17 illustrates the effect of summing two cosine waves and Fig. 7.18 the degree of complexity achieved by summing individual simple sine and cosine waves to produce synthetic Fourier series. The manner in which a complex curve can be filtered and broken down into a number of simple periodic components, some susceptible of rationalization in terms of process, is illustrated by the analysis of mean monthly precipitation (P) data for Madison, Wisconsin in Fig. 7.19. The annual trend of mean

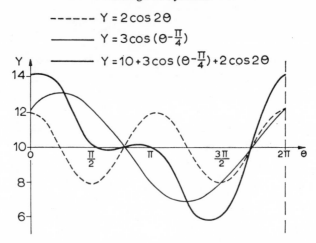

Fig. 7.17   The effect of the summation (i.e. interaction) of two
cosine curves. (After Rayner)

monthly figures is well described by the super-imposition of 6 series of sine
waves:

$$P = A_0 + A_1 \sin(30°t + \Phi_1) + A_2 \sin(60°t + \Phi_2)$$
$$+ \ldots + A_6 \sin(180°t + \Phi_6)$$

where: P = mean precipitation at month (t), as a deviation from the mean
(A_0).

A_0 = the arithmetic mean of the 12 monthly means (i.e. $P = O$).

A_1 to A_6 = the amplitudes of the six wave types, in terms of deviations from A_0.

t = the time, in months (i.e. Jan = 0, . . . Dec = 11).

Φ_1 to Φ_6 = the phase angles of the six wave types.

These curves exemplify a strong seasonal tendency (first harmonic) with a
July maximum (which is immediately obvious from the original data), a
strong double precipitation maximum in June and September (fourth
harmonic) (which is also obvious), and an interesting weak semi-annual
tendency with maxima in late April and late October (second harmonic)
which is not obvious from the original data plot. Thus the filtering of time
series, like that of spatial data referred to in Chapter, 4, can reveal hidden
tendencies in temporal data.

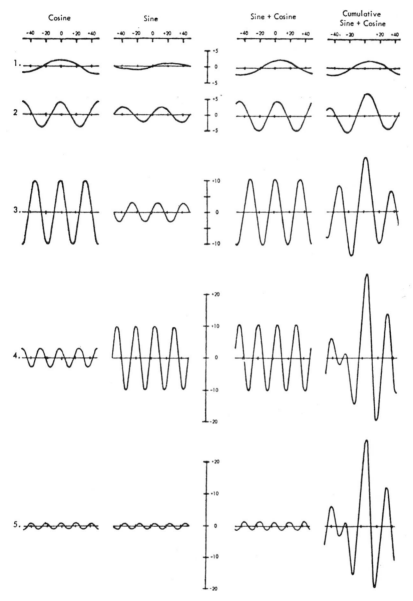

Fig. 7.18   Synthetic single Fourier series showing (*left*) the wave forms of the individual cosine and sine terms and (*right*) their paired summation (sine + cosine), together with a progressive summation of all the individual terms (cumulative sine + cosine). (From Harbaugh and Preston, 1967)

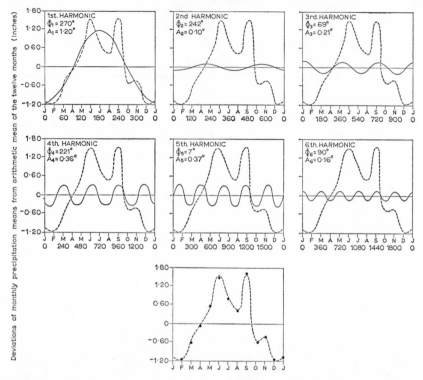

Fig. 7.19   The harmonic analysis of mean monthly precipitation data for Madison, Wisconsin. (After Horn and Bryson, 1960.) A description is given in the text. The circles in the bottom diagram were computed from the six harmonics. The dashed line is the curve of observed precipitation

The analysis of the components of systems change associated with so-called random noise presents special problems, particularly because what is commonly termed noise often includes subtle patterns of change which may be difficult to identify. An example of this can be given with reference to the data of Fig. 7.6. Application was made of a first-order Markov analysis to the residuals from the linear regression, where

$$R_{k+1} = r_{a1} \cdot R_k + e_{k+1}$$

in which: R_{k+1} = the residual which succeeds the residual R_k in the time series:

r_{a1} = the value of the autocorrelation coefficient for residuals which are one sampling time interval apart.

e_{k+1} = a random number.

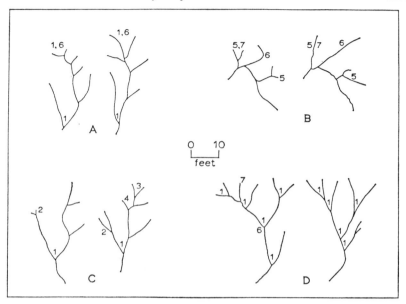

Fig. 7.20   Drainage-pattern changes in selected basins between 1948 and 1952 at Perth Amboy, New Jersey. Basins *a*, *c*, and *d* have steep gradient streams. Basin *b* is a youthful basin on the upper surface of the terrace. Drainage changes are indicated by numbers on the figures: 1. Angle of junction change. 2. Migration of junction. 3. Bifurcation. 4. Addition of tributary. 5. Angle of bifurcation change. 6. Channel straightening. 7. Elimination of tributary. (After Schumm, 1956)

In this instance the low value obtained for r_{a1} (0·24) made the possibility of autocorrelation (i.e. the existence of 'pattern' or 'non-randomness' among the residuals) very questionable. However, the fact that the 20th order polynomial curve provides a significantly better fit than that of the linear regression (at the 95% confidence level) does suggest that systematic variations in the residuals may be present.

In the foregoing treatment of the analysis of observed trends it is apparent that most attention has centred on temporal changes of system inputs, notably of precipitation and temperature. The resulting morphological changes within systems are commonly slower and more complex than those of individual input variables, and their observation consequently more difficult. In geomorphology morphological transformations are characteristically slow, but some records of observed changes have been made, particularly in small-scale slope and drainage systems on weak material. For example, Fig. 7.20 shows detailed drainage network changes which took place over a four-year period in an industrial sand and clay dump, and Fig. 7.21 the growth of meltwater stream channels on a glacier during a similar four-year interval.

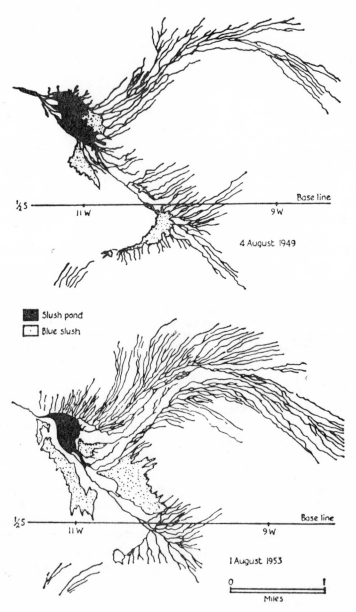

Fig. 7.21 Drainage pattern of meltwater streams on glacier, Shaw's Valley, Greenland Ice Cap. (*a*) August 4, 1949. (*b*) August 1, 1953. (After Holmes. From *Fluvial Processes in Geomorphology* by Luna B. Leopold, Gordon Wolman and John P. Miller. W. H. Freeman & Company. Copyright © 1964)

7.4 The Ergodic Hypothesis

Another method of attacking the problem of change in systems is by assuming that the statistical properties of a time series are essentially the same as the properties of a set of observations of the same phenomenon taken over a spatial ensemble. Although strict 'ergodicity' has rigorous statistical assumptions, the *ergodic hypothesis* suggests that, under certain circumstances, sampling in space can be equivalent to sampling through time and that space-time transformations are permissible. Clearly, not all morphometric features exhibit ergodicity, for example those relating to hydraulic geometry, since at-a-station changes through time differ from down-stream features at a given time. However, interesting results have been obtained by placing regional valley-side slope profiles and drainage networks in assumed time sequences. It must be clear that only very general hypotheses can be tested without absolute dates (i.e. by employing an ordinal, rather than an interval scale), and that the investigator always runs the risk of arguing in circles, by assuming that a temporal sequence exists because there are spatial variations when, in reality, the latter merely represent chance fluctuations around a timeless equilibrium state. Nevertheless spatial morphologic assemblages genuinely representative of temporal sequences have played a considerable part in the development of geomorphic theory.

Profiles surveyed on various residuals in the shale badlands of South Dakota have been placed in assumed time sequences based on decreasing height (Fig. 7.22), and the conclusion reached that, whereas the slopes of the less permeable Brule Formation tend to retreat parallel through time (probably by sheet wash), those of the more permeable Chadron Formation decline in angle as they are subjected to creep. Another study considered the development of slopes in a drainage basin which appeared to be actively extending into a river terrace, by the bifurcation of existing channels and the headward growth of fingertip tributaries. In this case it was felt justifiable to assume that, as there seemed to be a general tendency for all streams within the basin to increase in order as the net extended, one could use the order of any individual channel segment to indicate the relative age of the adjacent valley-side slope. Fig. 7.23 shows a plot of maximum valley-side slope angles for each of six stream orders, suggesting that initial channel incision is associated with the progressive steepening of the slope until this reaches a limiting angle close to its angle of repose. The size of the sample of highest order slopes was too small, and the order of the whole network was too low, to permit significant inferences regarding the later stages of slope development when, perhaps, decline in angle might have become important. A more recent study of a much larger and more 'mature' drainage basin in metamorphic rocks in Queensland has shown very much the same picture of

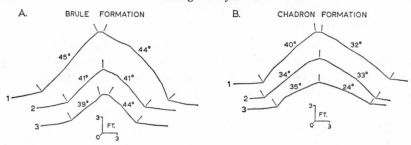

Fig. 7.22   Two series of slope profiles measured on residuals of (*a*) the Brule Formation and (*b*) the Chadron Formation, South Dakota. (After Schumm, 1956)

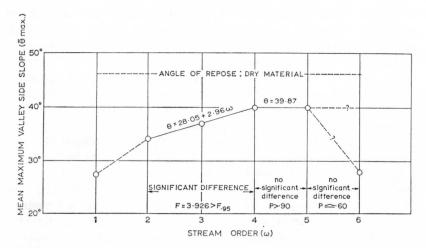

Fig. 7.23   Mean maximum valley-side slope angles versus stream order for a small drainage system actively developing in a terrace of the Farmington River, Connecticut. (From Carter and Chorley, 1961)

changing valley-side slope angles with increasing order. The indication that slopes decline along the highest-order valley sections also emerged in this study. An even clearer example of sequential slope development was provided by a study of the extent of subaerial denudation on an old sea cliff in South Wales. This cliff has been progressively protected from wave attack as the Laugharne Burrows marine spit (Fig. 7.24*a*) has extended eastwards and as a consequence the cliff profiles could be placed in a time sequence with respect to their relative protection from all but subaerial processes (Fig. 7.24*b*). Measurements of maximum slope angles of the recessional moraines of the Athabasca Glacier, Alberta, provided a still more precise time sequence, in that the

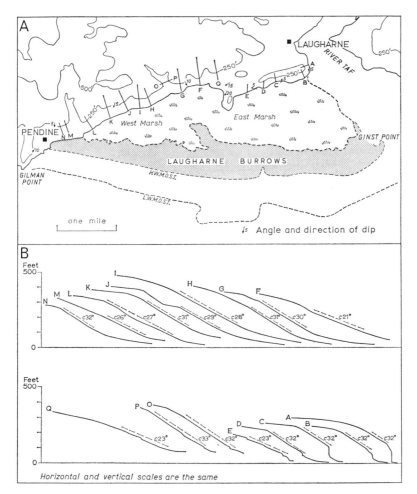

Fig. 7.24   The retreat of differentially-protected marine cliff slopes. (After Savigear, 1952)

(*a*) The cliffed coast between Pendine and Laugharne, South Wales, protected from marine erosion by the eastward-growing Laugharne Burrows. The locations of the cliff profiles depicted in (*b*) are shown.

(*b*) Profiles of the cliff sections located in (*a*) arranged so that the subaerially-youngest are on the right and the oldest on the left.

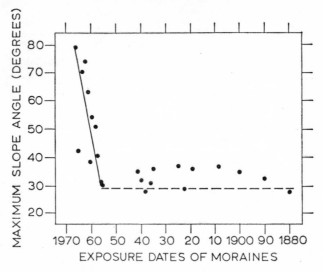

Fig. 7.25   Maximum recorded slope angles for the recessional moraines of the Athabasca Glacier, Alberta. (After Welch, 1970)

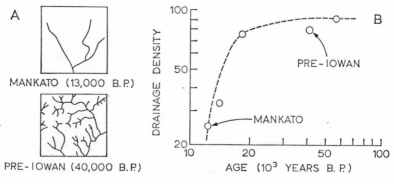

Fig. 7.26   Development of stream patterns on till sheets of different ages in the Des Moines lobe, central USA. (After Ruhe, 1952 and Leopold, Wolman and Miller, 1964. From Haggett and Chorley, 1969)

moraines could be more-or-less accurately dated by vegetational means. As Fig. 7.25 shows, the original steep collapse profiles (having quite a large standard deviation) rapidly declined to a maximum equilibrium angle of some 28° within less than 20 years. The value of such an ergodic sequence of spatial forms accurately calibrated with respect to time, as in the last example, is the degree of sophistication of the temporal model which can be tested, such that in this instance it is clear that there exist two slope development phases, each linear with respect to time, separated by a marked discontinuity.

Absolute dating of glacial deposits was similarly used in examining the drainage networks of differing ages which have developed on the glacial till sheets in Iowa. The difference was particularly well marked between the older Iowan and Tazewell, on the one hand, and the younger Cary and Mankato, on the other; and it was proposed that the drainage density had increased through time (as did the number of tributaries), rapidly for about the first 20,000 years and then more slowly (Fig. 7.26). One of the shortcomings of this study is that it failed to take fully into account the effect of other factors, notably differences in drift composition, which might affect drainage density independent of time. A most elaborate inferential investigation into the possible manner of evolution of stream networks was based on study of a large number of drainage basins in the western United States which assumed that they must have evolved in some way during time and therefore must represent, at this time instant, a considerable range of ages, so that this large basin sample must include representatives of many stages of development. However, there were apparent none of the systematic variations which one might expect if absolute age is associated with variations in drainage density (D), and it was assumed, not altogether surprisingly, that the wide range of climatic, soil and geological environments represented must also have affected the range of network forms observed. It is apparent that drainage density ($=$total length of streams (ΣL)/total basin area (a)) is distinct from stream frequency ($F =$ number of streams of all Strahler orders ((ΣN_u)/a). For sample areas which are much larger than their component first-order basins, D and F showed no systematic variations with A, but data for the mature basins (i.e. in which smooth valley-side slopes continue up to the divides) showed a systematic relationship between D and F, with a coefficient of correlation of $+0.97$, such that for all basins it was possible to propose that F/D^2 (the relative channel density) is a dimensionless constant (Fig. 7.27). The relative channel density therefore represents a basic law of behaviour of planimetric elements of maturely developed drainage basins and is a dimensionless measure of the completeness with which the channel net fills the basin outline for a given number of channel segments, and for any value of a. The function $F = 0.694 \, D^2$ thus represents a drainage network growth model, which shows that, as drainage density increases within a constant area, so does stream frequency.

7.5 Simulation Models

An important way of investigating the development of systems through time is by the use of simulation models. These are commonly of three types—scale, analogue and mathematical. Scale models use equivalent real world materials in a geometrically and kinematically scaled-down form (e.g. fine

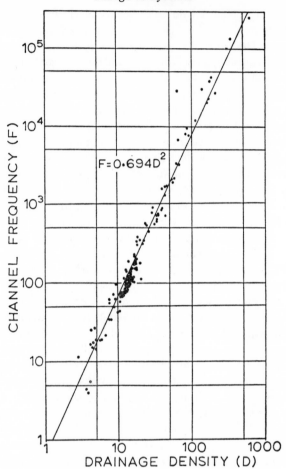

Fig. 7.27   Relation between channel frequency and drainage density. (After Melton, 1958. From Haggett and Chorley,
1969)

sand and water in a flume to simulate meander development), and analogue
models use different materials (e.g. in an electrical ground-water analogue
electrical potential may represent height of the water table and current the
flow of ground water). In general, the representations of real world systems
to scale or by analogy is accompanied by such intrinsic difficulties that
increasing attention is being paid to the exploration of time changes by means
of computer-based mathematical simulation models. Thus simulation is the
operation of an artificial model, which counterfeits some aspects of a real
world system, in such a way that the behaviour of the system is reproduced

as it moves through time. Examples are given here of mathematical simulation models of stream networks, valley-side slopes and deltas.

There are basically three types of stream network mathematical simulation models: (1) Growth models: which develop by headward extension and branching on an initially uneroded land surface. These have the disadvantage of predicting networks much more regular than are observed in nature, and tend to attain some static equilibrium when they have reached a given state. (2) Random models: which consider network processes to involve probabilistic choices leading to alternative final states. These give little insight into the processes or stages involved in network evolution, become static after generation, and produce excessive wanderings and near-loopings which are rare in nature. (3) Capture models: which start with an initial network and investigate its subsequent stability in the face of probabilities of capture and migration of divides. We have already encountered examples of the growth and random models in Horton's concept of drainage development (Chapter 6, Fig. 6.21) and the random-walk model (Chapter 6, Fig. 6.17a), respectively. A modification of the latter model has been developed by allowing biased networks (i.e. having greater probabilities of extending in certain directions), and by setting up a series of rules to prevent source junctions, triple junctions, looping and trapping. One programme was used to generate some 600 basins of third to fifth order (those with $N_1 \geqslant 10$ being selected for analysis), with a maximum drainage density (i.e. all matrix squares having a stream), which were divided into two classes: (1) Random: These were generated in a manner somewhat similar to the above model, using a 40×40 grid with equal probabilities of streams 'stepping' into each of the four adjacent squares (i.e. $P \downarrow = P \uparrow = P \leftarrow = P \rightarrow = 0.25$) (Fig. 7.28$a$); (2) Biased: These games were modified to employ a 40×60 grid, with four times the probability of a stream extending in one direction than in the opposite one (i.e. $P \downarrow = 0.4$; $P \uparrow = 0.1$; $P \leftarrow = P \rightarrow = 0.25$) (Fig. 7.28$b$). The relationships resulting from the biased games compared quite well with those of third-order natural stream networks, although the networks simulated by computer regularly departed from prediction as the number of first-order streams decreased. The random models thus illustrate one of the major difficulties in employing mathematical simulation models to explore system changes through time, namely that the model may have to be so simplified and its development may proceed through such unreal and artificial steps that its only value lies, not in its evolutionary representations, but in its ability to approach and exemplify some equilibrium state.

It is thus clear that such simulation models may differ from other evolutionary approaches in that, by simulating the end result of channel network evolution, it is hoped to throw light on some general principles governing evolution, rather than to simulate the stage-by-stage development. Such

20

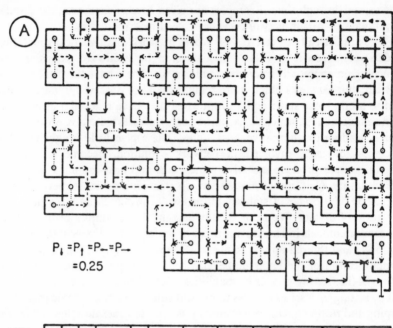

$$P_\downarrow = P_\uparrow = P_\leftarrow = P_\rightarrow$$
$$= 0.25$$

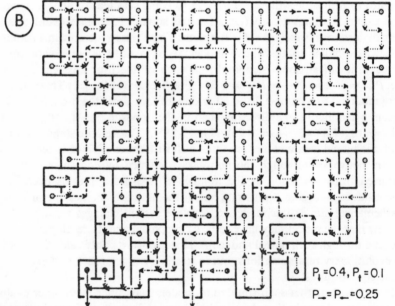

$$P_\downarrow = 0.4, P_\uparrow = 0.1$$
$$P_\leftarrow = P_\rightarrow = 0.25$$

Fig. 7.28   Portions of channel networks generated by computer programme: (*a*) random game, (*b*) biased game, as described in the text. (After Smart, Surkan and Considine. From Haggett and Chorley, 1969)

studies seem to indicate that although every stream junction has its own *raison d'être*, the mechanics of these small-scale effects is so complicated that the details concerning most individual events will remain forever unknown. Thus the areal variability of process can never be completely explained, but the end result is as if the process was largely random. Nevertheless, it must be recognized that such randomly-generated networks do *not* simulate the actual geomorphic processes involved in the evolution of natural networks, which processes usually take place simultaneously over an exposed area. This means that, although mathematical models provide fundamental general principles from which to begin the study of drainage basins, neither the growth nor the random-walk descriptions necessarily indicate how the nets develop and change through time. An attempt to circumvent this difficulty is represented by the capture model, one of which contained 450 squares in an 18 × 25 matrix and assumed that parallel streams imparted a uniform drainage density and that there was a single drainage outlet (Fig. 7.29).† Certain initial assumptions were made, namely that the relative elevations were as shown in Fig. 7.29(1), and that the probability of capture of a given stream segment by one in an adjacent square was a function of their upstream drainage areas (which control discharge and the amplitude of adjacent meanders), and of gradient across the site of potential capture. The operation of the capture simulation model involves the following steps: (1) Randomly select a point on the stream matrix; (2) Examine the four surrounding squares in terms of the probability of their capturing the stream selected; (3) If capture is allowed, regrade the upstream drainage area of the captured stream according to the formula empirically derived from field data: stream gradient = contributing basin area$^{-0.6}$; (4) Select another random point of possible capture, and so on, until the desired number of trials or captures has occurred. Fig. 7.29 (2–4) shows one series of simulations after 4, 100 and 319 captures, respectively. Repeated runs showed, firstly, that network changes increased rapidly to a peak and then declined to a fairly constant value, and, secondly, that when the resulting networks were ordered according to the Strahler system they agreed quite well with natural networks. From the above preliminary analysis, it is suggested that capture may be a significant process in the development of some natural stream networks, for example to stream systems on pediments.

A similar computer-based fluvial erosion simulation model has been developed‡ on the basis of a set of 30 × 30 initial altitudes whose orthogonal

† A. D. Howard of the Department of Geography and Environmental Engineering, the Johns Hopkins University (mimeo) gave kind permission to use this unpublished material.

‡ B. Sprunt of the Department of Geography, Portsmouth Polytechnic (personal communication) gave kind permission to use this unpublished material.

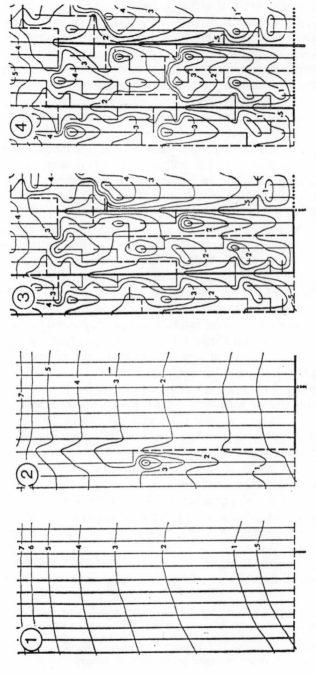

Fig. 7.29 Stages in Howard's capture model simulation, after zero (1), four (2), one hundred (3), and 319 captures (4). (After Howard. From Haggett and Chorley, 1969)

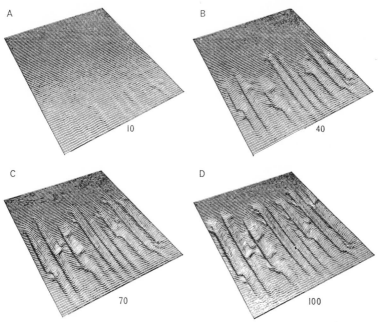

Fig. 7.30   Simulations number 10, 40, 70 and 100 in the development of a system of drainage basins on an inclined plane simulated by digital computer. (From unpublished material kindly made available by Brian Sprunt, Portsmouth Polytechnic.) A fuller description is given in the text

projection on a horizontal plane produces a regular lattice of points which are used to define the drainage network in the form of a non-planar directed graph. In the simulation depicted in Fig. 7.30 the initial surface was an inclined plane having a grid spacing of 250 feet and a maximum relief of 900 feet. Peak point runoff could be determined by assuming regularly-repeated inputs of runoff at each point. Runoff inputs were generated by distributing 900 units over the 30 × 30 points using 900 random pairs of co-ordinates, this being the only stochastic element in the operation of the model. When runoff was allowed in more than one direction from a point it was distributed in proportion to the sines of the slope angles in each direction; no infiltration losses of runoff were allowed; peak runoff at each point was computed by simply carrying flows downhill from sources to outlets of the model; and flows crossing the model boundaries were considered lost. Stream lines were arbitrarily defined as possessing flows of 6 + 40/(the number of the simulation), so allowing stream lengths to rapidly increase during the early simulations and then to approach an asymptotic value. Erosion at each point was automatically calculated at regular simulation time intervals and was assumed to vary as (1) the peak runoff determined for the point, (2)

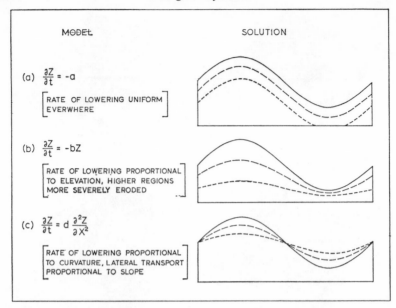

MODEL

SOLUTION

(a) $\frac{\partial Z}{\partial t} = -a$

$\begin{bmatrix} \text{RATE OF LOWERING UNIFORM} \\ \text{EVERWHERE} \end{bmatrix}$

(b) $\frac{\partial Z}{\partial t} = -bZ$

$\begin{bmatrix} \text{RATE OF LOWERING PROPORTIONAL} \\ \text{TO ELEVATION, HIGHER REGIONS} \\ \text{MORE SEVERELY ERODED} \end{bmatrix}$

(c) $\frac{\partial Z}{\partial t} = d\frac{\partial^2 Z}{\partial X^2}$

$\begin{bmatrix} \text{RATE OF LOWERING PROPORTIONAL} \\ \text{TO CURVATURE, LATERAL TRANSPORT} \\ \text{PROPORTIONAL TO SLOPE} \end{bmatrix}$

Fig. 7.31   Simple deterministic models of the progressive modification of topography having a sinusoidal profile, assuming:

(*a*) A uniform rate of lowering.
(*b*) A rate of lowering proportional to elevation.
(*c*) A rate of lowering proportional to curvature.
(After Pollack, 1969)

a theoretical function of slope derived from Horton, and (3) the maximum available relief between the point and its neighbours. After each simulation the amount of erosion was subtracted from the previous elevation of each point such that the surface of the model was changed by successive simulations and these, in turn, determined the topology of the stream network. This topology was also controlled by the amount of flow through a point when all points were experiencing peak runoff. Below a certain level flow was diffused from each point and the network was non-planar, above this level runoff was assumed to take place only in the direction of steepest slope from each point, producing a directed stream network. Figure 7.30 shows simulations number 10, 40, 70 and 100, respectively, in the progressive degradation of the initial sloping plane surface through time.

Another distinction between different types of mathematical simulation studies is exemplified by the following valley-side slope models, i.e. that which exists between deterministic and stochastic simulation models of time transformations. Deterministic mathematical models are based on classical mathematical notions of predictable relationships between variables (i.e.

between cause and effect), and consist of a set of exactly specified mathematical assertions (derived from experience, theory or intuition) from which unique consequences can be derived by logical mathematical argumentation. Such models have been applied to the problem of valley-side slope development for over 100 years and a very simple example involving a sinusoidal profile is given in Fig. 7.31. Such a model involves, firstly, the accurate specification of some realistic natural form; secondly, the assumption of the total influence of composite erosion and deposition along the profile; thirdly, its mathematical expression and the carrying out of the resulting changes through time mathematically; and, fourthly, the testing of the resulting forms against the real world. Of these, steps two and four are the most difficult, the former because of the difficulty of anticipating the composite spatial effect of a group of interlocking processes, and the latter because of the difficulty of testing the significance of difference between the geometrical results of simulation models operated under very different assumptions. Indeed, few deterministic statements can completely specify all the significant variables included in a given complex natural situation, so that discrepancies occur which, together with the random unpredictable effects inherent in all natural processes, combine to produce a noise level or variance which tends to obscure the simpler deterministic relationships.

A very effective approach to such deterministic relationships has been recently made in connection with transportational slopes of many kinds by reference to the transport law:

$$C = f(a)\left(-\frac{\partial y}{\partial x}\right)^n$$

where C = transport capacity; a = area 'drained' per contour length ($f(a)$ thus describes the influence of increasing distance from the divide); x = horizontal distance from the divide; y = elevation above baselevel; n = a constant exponent (zero or positive) describing the influence of increasing gradient.

From this equation can be derived the simple empirical transport capacity relationship:

$$C \propto a^m . (\text{slope})^n$$

The following values for m and n have been obtained from field measurement of forms associated with different transportational processes:

	m	n
Soil creep	0	1·0
Rainsplash	0	1·0–2·0
Soil wash	1·3–1·7	1·3–2·0
Rivers	2·0–3·0	3·0

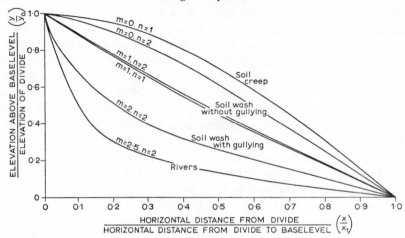

Fig. 7.32   Dimensionless graph showing approximate characteristic-form slope profiles for a range of transportational processes. (After Kirkby, 1971)

From these relationships a family of dimensionless curves can be generated characteristic of a wide range of transportational processes (Fig. 7.32). It is interesting that, by doubling the length of the Y (y/y_o) axis in Fig. 7.32, we obtain the dimensionless curve of the height/length integral (Fig. 2.21). Consequently, for *simple* slopes, it may prove possible to estimate m and n (and, as a result, the dominant transportational process) directly from the height/length integral. However, the latter index is rather insensitive in the range 40–60%, which represents the critical transition zone between concentrated and unconcentrated wash ($m = 2, n = 2; m = 0, n = 2$); furthermore, most slopes for which the height/length integral has been calculated as yet are far from simple in form. Nevertheless, it does appear that the functional importance of this purely morphological index may be substantial in some cases.

Often, however, the random effects are so important in influencing the results of natural process that dominantly statistical (stochastic) models have to be constructed to take account of them. For example, a sophisticated model of three-dimensional stochastic movement of soil particles on relatively gentle humid slopes has been developed, in which gravity and gravitational soil moisture movements provide a downward bias which is opposed by the decrease of soil pore space with depth (i.e. of increasing packing and density). Where no surface gradient exists the model predicts a density layering parallel to the surface, whereas with a gradient a slow mean resultant downslope movement of soil particles occurs, depending on the density gradient, in response to a tendency to replace layering parallel to the surface by horizontal layering.

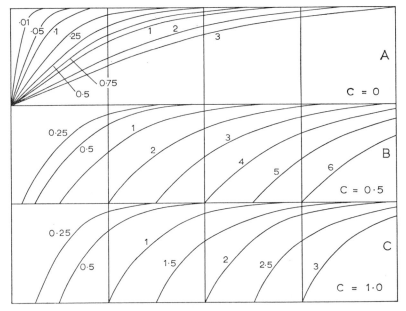

Fig. 7.33 Profiles resulting from the operation of a three-dimensional stochastic model of soil creep on humid slopes, assuming three different strengths (c) of lateral stream undercutting. A more complete description is given in the text. (After Culling, 1963)

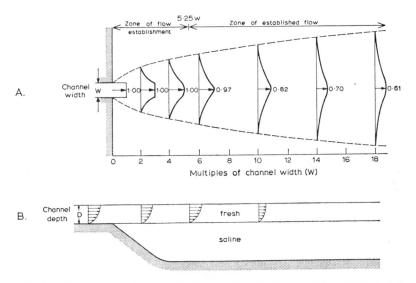

Fig. 7.34 The structure of the velocity field employed in the simulation of delta building. (After Bonham-Carter and Sutherland, 1968 and Harbaugh and Merriam, 1968)

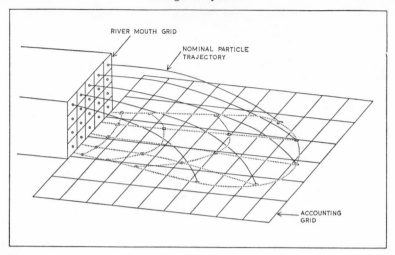

Fig. 7.35   The structure of a delta-building simulation model, showing the vertical river mouth grid and the horizontal accounting grid. The model allows statistical particles issuing from the centre of each river mouth grid cell to move along calculated trajectories until they settle into the accounting grid cells. (From Bonham-Carter and Harbaugh, 1968)

This dominantly stochastic model takes into account the effects of basal stream action as providing an absorptive boundary for the downslope diffusion of soil particles, and the model can operate differentially as the result of variations in the strength of lateral undercutting (c) by the basal stream (Fig. 7.33).

A final example of a partly deterministic, partly stochastic, mathematical simulation model is provided by one developed to represent the delta-building process. The horizontal and vertical structures of the velocity field of a plane jet flowing into a body of standing water were first assumed (Fig. 7.34), and then a river mouth grid was constructed in which every cell possessed a different nominal particle trajectory, depending on particle grain size, stream velocity structure, slope, depth, etc., variations of which lead to different dispositions of grain deposits to be built up on the accounting grid (Fig. 7.35). The model contained a feedback by placing a limit on the number of particles which could be accumulated in each box of the accounting grid, after which there was a stochastic reallocation of the excess particles. One series of 'runs' of the model, for example, assumed a discharge of 2·85 m/sec, a stream cross-section of 1 m × 20 m, a stream gradient of 17 in 1000, leading to a deposit of 40% porosity on a flat original bottom geometry, and produced three depositional patterns for the particle sizes (0·3 mm, 0·2 mm and 0·1 mm, respectively) (Fig. 7.36). Fig. 7.37 shows,

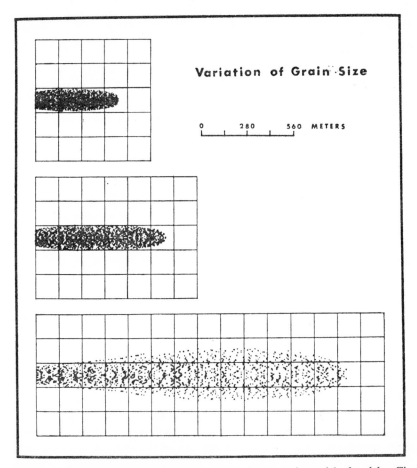

Fig. 7.36   Results from computer experiments using a static model of a delta. Fine points are terminal coordinates of statistical particles as they settled into horizontal accounting grid cells. River channel dimensions are 100 metres wide and 10 metres deep, with average current velocity of 2 metres/sec. The three diagrams record sensitivity of the model to grain-size changes. Sediment diameters: 0·3 mm (*upper*), 0·2 mm (*centre*), 0·1 mm (*lower*). (From Bonham-Carter and Harbaugh, 1968)

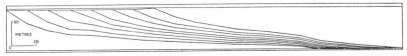

Fig. 7.37   Longitudinal section through a simulated deltaic complex showing the geometric configuration of deposits formed during a simulation run involving eight time increments. (After Bonham-Carter and Sutherland. From Harbaugh and Merriam, 1968)

finally, the predicted development through time of the longitudinal section
of a deltaic complex produced by such a model through eight time increments.

7.6 Conclusion

It is clear that, despite the swing in emphasis from 'timebound' to 'timeless'
studies of systems in physical geography, the problem of change through time
remains absolutely central to our understanding of the landscape. We need
perhaps to re-emphasize the basic difficulties involved.

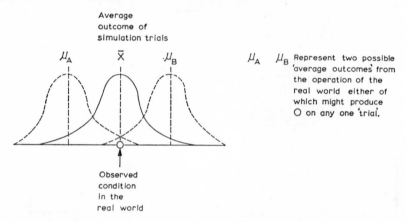

Fig. 7.38   A diagrammatic illustration of one of the main sources of incertainty in the
interpretation of simulation studies

Firstly, there is the problem of *equifinality*: that similar changes may result
from totally disparate combinations of input, throughput and output acting
over widely varied periods of time. This problem is particularly thorny when
simulation models are used to attempt to explain changes in an individual
landscape as, although the model may be run many times to give an average
condition which fits that observed in the real world, the latter case represents
the outcome of a *single* evolutionary process and, in statistical terms, is
quite likely to belong to a very different population (Fig. 7.38). An allied
problem arises from the ability of two simulation models with quite different
constraints to produce outcomes that are virtually indistinguishable and both
closely akin to reality. This difficulty is particularly apparent with respect
to simulations of river long profiles.

Secondly, we have the problem of deciding what constitutes 'change'.
Although this basically relates to the question of *time* scale, it is also bound
up with that of *measurement* scale: in other words, how accurately can we
perceive variations in each aspect of the phenomenon under study? Allied
to this is the whole question of the differing statistical properties of the many

classes of variables of interest to the physical geographer. By and large we are able to detect change most accurately in those variables (e.g. valley-side slope angle) which may be measured on the interval scale and are normally distributed, and least certainly in those (e.g. degree of rock weathering) which may only be measured on some arbitrary, ordinal scale and consequently follow no fixed statistical distribution.

Finally, there is the persistent and grave likelihood—inherent in the use of the ergodic assumption—that we may employ circular argument and wrongly equate difference in form with difference in stage of evolution.

For all these problems, at present, the best solution available is the exercise of rigid, common-sense restraint on the part of the physical geographer in his interpretation of change within any system in the landscape. It is to be hoped that refinement of statistical techniques will help to solve some of these dilemmas, but even so this can never replace intelligent questioning of the *likelihood* of results.

REFERENCES

Agterberg, F. P. (1967), Computer techniques in geology; *Earth Science Reviews*, **3**, 47–77.

Ahnert, F. (1970), Functional relationships between denudation, relief, and uplift in large mid-latitude drainage basins; *American Journal of Science*, **268**, 243–263.

Arnett, R. R. (1971), Slope form and geomorphological process: An Australian example; In Brunsden, D. (Ed.), *Slopes:Form and Process*, Institute of British Geographers, Special Publication No. 3, 81–92.

Barry, R. G. and Chorley, R. J. (1971), *Atmosphere, Weather and Climate*, 2nd Edn. (Methuen, London), 379.

Barry, R. G. (1969), Long term precipitation trends; In Chorley, R. J. (Ed.), *Water, Earth and Man* (Methuen, London), 513–523.

Bonham-Carter, G. and Harbaugh, J. W. (1968), Simulation of geologic systems: An overview; *State Geological Survey, University of Kansas, Computer Contribution 22*, 3–10.

Bonham-Carter, G. and Sutherland, A. J. (1968), Mathematical model and Fortran IV program for computer simulation of deltaic sedimentation; *State Geological Survey, University of Kansas, Computer Contribution No. 24*, 56.

Carter, C. S. and Chorley, R. J. (1961), Early slope development in an expanding stream system; *Geological Magazine*, **98**, 117–130.

Chorley, R. J. (1962), Geomorphology and general systems theory; *U.S. Geological Survey, Professional Paper 500-B*, 10.

Chorley, R. J. (1967), Models in geomorphology; In Chorley, R. J. and Haggett, P. (Eds.), *Models in Geography* (Methuen, London), 57–96.

Connelly, D. S. (1968), *The Coding and Storage of Terrain Height Data: An introduction to numerical cartography* (M.Sc. Thesis, Cornell University), 140.

Conrad, V. and Pollack, L. W. (1950), *Methods in Climatology* (Harvard), 459.

Croxton, F. E. and Cowden, D. J. (1948), *Applied General Statistics* (Prentice Hall, New York), 944.

Curry, L. (1962), Climatic change as a random series; *Annals of the Association of American Geographers*, **52**, 21–31.

Curry, L. (1966), Chance and landscape; In House, J. W. (Ed.), *Northern Geographical Essays* (Oriel Press, Newcastle upon Tyne), 40–55.

Derdariani, A. S. (1967), A plane mathematical model of the growth and erosion of an uplift; *Soviet Geography*, **8**, 183–198.

Dorf, E. (1960), Climatic changes of the past and present; *American Scientist*, **48**, 341–364.

Fischer, R. (Ed.), (1967), Interdisciplinary perspectives of time; *Annals of the New York Academy of Science*, **138**, Art. 2, 367–915.

Gerard, R. W. (1964), Entitation, animorgs, and other systems; In Mesarović, M. D. (Ed.), *Views on General Systems Theory* (Wiley, New York), 119–124.

Haggett, P. and Chorley, R. J. (1969), *Network Analysis in Geography* (Arnold, London), 348.

Harbaugh, J. W. and Bonham-Carter, G., (1970), *Computer Simulation in Geology* (Wiley-Interscience, New York), 575.

Harbaugh, J. W. and Preston, F. W. (1968), Fourier series analysis in geology; In Berry, B. J. L. and Marble, D. F. (Eds.), *Spatial Analysis: A reader in statistical geography* (Prentice Hall, New Jersey), 218–238.

Harbaugh, J. W. and Merriam, D. F. (1968), *Computer Applications in Stratigraphic Analysis* (Wiley, New York), 282.

Harvey, D. (1969), *Explanation in Geography* (Arnold, London), 521.

Holloway, J. L. (1958), Smoothing and filtering of time series and space fields; *Advances in Geophysics*, **4**, 351–389.

Horn, L. H. and Bryson, R. A. (1960), Harmonic analysis of the annual march of precipitation over the United States; *Annals of the Association of American Geographers*, **50**, 157–171.

Howard, A. D. (1965), Geomorphological systems—Equilibrium and dynamics; *American Journal of Science*, **263**, 302–312.

Kennedy, B. A. (1965), *An Analysis of the Factors Influencing Slope Development on the Charmouthien Limestone of the Plateau de Bassigny, Haute Marne, France* (Unpublished B.A. Dissertation, Cambridge University).

Kirkby, M. J. (1971), Hillslope process-response models based on the continuity equation; In Brunsden, D. (Ed.), *Slopes: Form and Process*, Institute of British Geographers, Special Publication No. 3, 15–30.

Leopold, L. B. (1951), Rainfall frequency: An aspect of climatic variation; *Transactions of the American Geophysical Union*, **32**, 347–357.

Leopold, L. B., Wolman, M. G. and Miller, J. P., (1964), *Fluvial Processes in Geomorphology* (Freeman, San Francisco), 522.

Lewis, P. (1960), The use of moving averages in the analysis of time-series; *Weather*, **15**, 121–126.

Marujama, M. (1963), The second cybernetics: Direction-amplifying mutual causal processes; *American Scientist*, **51**, 164–179.

Merriam, D. F. (Ed.) (1969), *Computer Applications in the Earth Sciences* (Plenum Press, N.Y.), 281.

Miller, R. L. and Kahn, J. S. (1962), *Statistical Analysis in the Geological Sciences* (Wiley, New York), 483.

Moultrie, W. (1970), Systems, computer simulation and drainage basins; *Bulletin of the Illinois Geographical Society*, **12**(2), 29–35.

Pollack, H. N. (1969), A numerical model of the Grand Canyon; *In Geology and Natural History of the Grand Canyon Region*, Four Corners Geological Society Guidebook to the Fifth Field Conference, Edited by Baars, D. C., 61–62.

Savigear, R. A. G. (1952), Some observations on slope development in South Wales; *Transactions of the Institute of British Geographers*, **18**, 31–51.

Scheidegger, A. E. (1970), Stochastic models in hydrology; *Water Resources Research*, **6**, 750–755.

Schumm, S. A. (1956), The role of creep and rainwash on the retreat of badland slopes; *American Journal of Science*, **254**, 693–706.

Schumm, S. A. (1963), The disparity between present rates of denudation and orogeny; *U.S. Geological Society, Professional Paper 454-H*, 13.

Schumm, S. A. and Lichty, R. W. (1965), Time, space and causality in geomorphology; *American Journal of Science*, **263**, 110–119.

Strahler, A. N. (1950), Equilibrium theory of erosional slopes, approached by frequency distribution analysis; *American Journal of Science*, **248**, 673–696 and 800–814.

Thom, R. (1968), Topological models in biology; *Topology*, **7**, 313–335.

Tobler, W. R. (1969), Geographical filters and their inverses; *Geographical Analysis*, **1**, 234–253.

Towler, J. E. (1969), *A Comparative Analysis of Scree Systems developed on the Skiddaw Slates and Borrowdale Volcanic Series of the English Lake District* (Unpublished B.A. Dissertation, Cambridge University).

Welch, D. M. (1970), Substitution of space for time in a study of slope development; *Journal of Geology*, **78**, 234–238.

8: *Control Systems*

So Man, who here seems principal alone,
Perhaps acts second to some sphere unknown,
Touches some wheel, or verges to some goal;
'T is but a part we see, and not a whole.

A. POPE: *An Essay on Man*, Epistle 1

8.1 Control Systems in Physical Geography

The emphasis on man's role in intervening in natural spatial process-response systems to change the face of the earth has, one way or another, been a focus of geographical interest for more than a century. However, much of the emphasis has been placed on man's unwitting or perverse degradation of physical and biological process-response systems by unplanned urban and industrial growth and by the destruction of 'natural' plant and animal relationships. Although there is some current emphasis on the recognition of conditions promoting systems stability or equilibrium so that they may be managed by small inputs of energy, man's relation to natural systems has consistently been confused by his many visions of this relationship. Deterministic notions stressed man's subordination to the natural world; teleological theories, his superiority to it; ecological studies, his integration with it; perceptual theory, his psychological bondage to it; economic theory, his mastery over increasingly large parts of it; game theory, his competition with it; liberal-humanist attitudes, his 'alien' influence on it, commonly manifested as a malevolent force 'upsetting the balance of nature', by forcing natural systems to pass across thresholds and so disrupting former equilibrium relationships. All these views possess an element of truth, but it is also true that man, organised in ever more effective decision-making groups, is increasingly able (at least in theory) to exploit his growing knowledge of the nature and operation of natural process-response systems so as to be able to intervene in them to exert an influence which will modify their operation in a planned and predictable manner, beneficial to man in the widest sense, without the concomitant occurrence of unforeseen injurious secondary effects, and without either destroying the system or causing its degradation by impelling it through critical operational thresholds.

The general characteristics of control systems have been outlined in Chapter 1.3. We have particularly stressed that key variables (or 'valves') can be identified in process-response systems, and that, at these points,

298

intelligence can most effectively intervene to produce operational changes in the mass/energy distribution and morphological relationships within the systems. This concept was foreshadowed long ago by J. Clerk Maxwell's identification of *singular points* within complex systems, at which a comparatively small applied force can effect relatively large transformations because of the character of the point and its position relative to the whole. As we have seen, regulators and storage variables are particularly effective principal points or valves in physical process-response systems (Fig. 8.1), although the processes of decision-making leading up to intervention are extremely complex (Fig. 8.2). Such a linkage of physical and socio-economic systems involves great difficulties because of at least two vital differences in their individual character:

1 Human and higher biological systems differ from physical systems in that they possess a memory of some kind which exerts some operational control.

2 As has been previously pointed out, the operation of physical systems is dominated by a tendency for negative feedback, whereas socio-economic systems possess strong positive-feedback loops which make change on-going.

It has been necessary to make constant reference to cybernetic ideas throughout this book in that cybernetics is the science of control in both the living organism and the machine, concerned with the flow of information in the system and its use as a means of control. From this point of view man-environment systems might be considered as a special case of man-machine systems where the machine contains a large number of black, grey and white box subsystems.

Although there are obvious differences between the concepts of regulation, influence and control as applied to the relationships between man in society and physical systems, it is usual to consider control as forming part of the feedback loop. Ashby's *law of requisite variety* states that for an intelligence to gain control over a system it must be able to take at least as many distinct actions as the system can exhibit. In other words, it must be able to take a sufficient number of counter measures so as to force down the variety of possible outcomes to within a restricted range commensurate with the objectives of the controller. This law teaches us that, in order to control a system, one must either increase the possible variety of actions available to one, or operate in some manner so as to restrict the variety of responses which the system can produce.

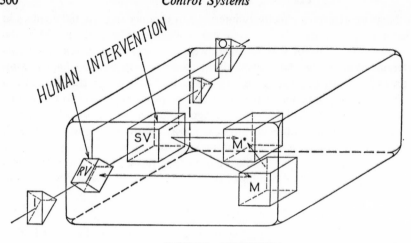

CONTROL SUBSYSTEM

Fig. 8.1   A control system illustrated by part of the process-response system shown in Fig. 1.3*c*, in which human intervention is operating on the regulator and the storage components as 'valves' (RV and SV, respectively). (From Chorley, 1971)

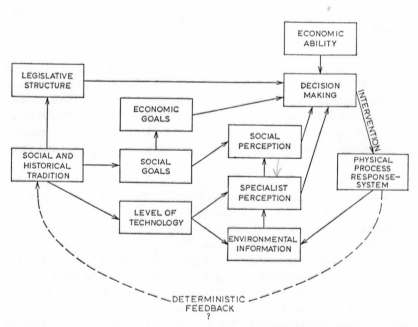

Fig. 8.2   Some schematic elements of a decision-making system, showing on the right the human intervention in the process-response system illustrated in Fig. 8.1. (From Chorley, 1971)

There are three types of feedback control which can be employed to regulate systems:

1 Goal-seeking—i.e., the type of negative feedback or homeostasis which was treated in Chapter 6.
2 Learning—in which *external* information enters into the feedback loop, having the power to alter the operation of the system.
3 Consciousness—in which information is generated *within* the system (i.e. there is self-awareness) and this enters as well into the feedback loop to produce changes in the nature of the system (i.e. on-going positive feedback).

With these ideas in mind, it has been possible to illustrate a hierarchy of feedback control systems which goes some way towards representing these types. Fig. 8.3. depicts the first four orders of such a six-tiered hierarchy, where:

0 Order is a simple input/output structural system with no feedback and no memory (Fig. 8.3*a*). A cynic has compared this system with private soldiers!
1 Order is a similar input/output cascade, possessing, however, a simple deterministic feedback; there is no memory (Fig. 8.3*b*). These are the Corporals of the control systems! The input or environmental vector (U_t) operates on the structural system at time t, produces an output (S_{t+i}) at time $t + 1$ which, in turn, produces a structural state vector which feeds back to modify the structural system.
2 Order is the sequential decision system possessing the loop of the decision feedback vector (Fig. 8.3*c*) within which the decision maker can operate in accordance with his own information regarding the system environment (i.e. environmental vector). At this order of system complexity, stochastic environmental influences are brought to bear on the system through the action of the decision maker. This second order might characterize the relations of a Sergeant to his squad.
3 Order is the adaptive decision system which, in addition to the 2 Order characteristics, possesses a historical information vector loop containing a record of past environmental observations by the decision maker in the form of stochastic data providing probability distributions as guides for the decision function (Fig. 8.3*d*). Those who have been army officers would cast the Company Commander in such a decision-making role.
4 Order is a system of greater complexity and memory storage which can learn to the extent of developing new predictions, rules and structural alterations so that the system can operate more efficiently towards a given

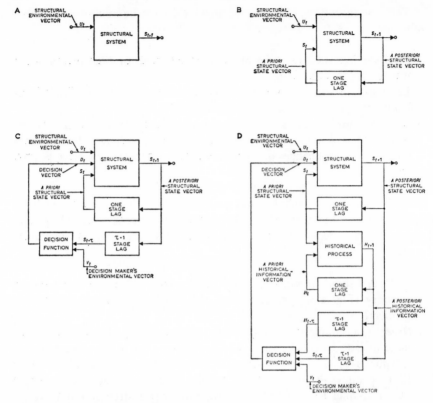

Fig. 8.3   The first four orders of a hierarchy of feedback control systems. (From Murphy 1965)

    (*a*) 0 Order. Simple input/output system.
    (*b*) 1 Order. Input/output system with simple deterministic feedback.
    (*c*) 2 Order. A sequential decision system possessing a decision feedback loop.
    (*d*) 3 Order. An adaptive decision system possessing an historical information feedback loop.

goal. A General and his Division might, charitably, be given as an example of such a 4 Order system.

5 Order is a system of still greater complexity, rich in feedback loops and monitoring of subsystems, with a highly structural memory, that can consciously direct itself towards the solution of new goals. An Army with its General Staff was once thought to represent such a system. In some respects, geographical control systems are of this order.

It is clear, that, when playing a role in the above systems, the decision maker has, variously, to be able to identify or determine his objective, the system variables, the stochastic elements, the historical information vector, the strategy for handling risk or uncertainty, the conditions of optimality, and the sensitivity of the system to his actions. The following sequence of evolutionary levels for living control systems given by Buckley shows how high on the ladder of complexity is situated the kind of decision-making upon which the operation of geographical control systems relies:

1 Genetic mutation and selective survival.
2 Bisexuality, allowing more rapid adjustment to environment changes.
3 Blind trial and error (i.e. homeostasis).
4 Learning. The retention of adaptive response patterns.
5 Perception. The visual exploration of potential behaviour alternatives.
6 Observational learning and imitation. From observations of other exploring consciousnesses.
7 Linguistic instruction. The transference of information about the environment and of correct responses to it.
8 Thought. The symbolic rehearsal of potential behaviour against a learned model of the environment.
9 Social decision-making. The pooling of ideas into a single environmental model.

8.2 Environmental Perception and Decision-Making

Much research is currently being directed towards the analysis of the decision-making systems outlined in Fig. 8.2. This work involves much study of the psychology, logistics, economics and sociology of individual and group decision making, but, at rock bottom, it is the attitude of the individual and of society to the natural environment which motivates and controls human environmental intervention. This attitude, of course, stems from a whole mass of social, technological and historical influences, and it is interesting that studies of environmental perception are already giving promise of an increasingly secure place for historical geography within contemporary developments of the geographical discipline as a whole, to a greater extent than has been possible for the past twenty years or so. It is clear also that the environment itself conditions much of man's attitude to it, although not in the classical deterministic sense.

Man's attitude to the natural environment seldom, if ever, involves an optimum response to its conditions. This means that in his dealings with the environment, whether as an individual or as a member of a group, man does not achieve the economic ideal of *optimization*, but tends to aim, instead, at *satisficer* behaviour. There are two main reasons for this departure from 'optimality' in human dealings with the environment:

1 There are a great number of things which man could choose to optimize: output/input value ratios on the short or the long term; security from hazard; the 'quality' of life or of landscape, and so on. Many of these are ideals—and vary in interpretation from one man or one society to another. In addition, one would frequently desire an optimization of several of these ideals simultaneously. This last would be mathematically impossible, even if one could define the 'optimum' in each ideal satisfactorily.
2 The attainment of the optimum even in a relatively simple case like output/input measures, generally requires a very much larger input of energy, in some form or other, than does the achievement of some slightly sub-optimum state. In other words, the law of diminishing returns operates.

TABLE 8.1

Estimates of average annual losses from selected geophysical hazards in the United States. Single year estimates are the level of average losses current to year cited. Property damage figures are in millions of dollars unadjusted unless otherwise noted.

HAZARD	Loss of Life		Annual or Average Annual Property Damage	
	No.	Period	Amount ($ million)	Period
Floods	70	1955–64	1000	1966
			350–1000	1964
			290	1955–64
Hurricanes	110	1915–64	250–500*	1966
			100	1964
			89	1915–64
Tornadoes	194	1916–64	100–200*	1966
			40	1944–64
			300	1967
Hail, Wind and Thunderstorms			125–250*	1966
			53	1944–53
Lightning strikes and Fire	160	1953–63	100	1965
Earthquakes	3	1945–64	15	1945–64
Tsunamis	18	1945–64	9	1945–64
Heat and Insolation	238	1955–64		
Cold	313	1955–64		
TOTALS	1106		621–2174	

*Insured losses only.

(From Burton, Kates and White, 1968).

For these reasons, most environmental decision making involves the selection of *some* alternatives, but not the selection of *optimum* ones. Because the processes of optimization are several times more complex than those required to satisfice, the true character of the former is being explored by computer simulation techniques. This involves massive data collection, the generation of synthetic data, complicated analyses involving linear and non-linear programming, and the construction of complex mathematical models which must be tested in the face of a mass of stochastic inputs. Chapter 8.5 explores this topic further.

Much of the work which has been directed towards man's attitudes and responses to his environment has been concerned with chance-laden hydro-meteorological events, particularly with floodplains and storm coastlines, and with drought-liable agricultural areas. There is the apparent paradox that, despite his technological power and increasing ecological dominance, man's vulnerability to environmental hazards seems to be growing (Table 8.1). The reasons for this are that technological manipulation of the environment, while reducing some hazards, produces others and that, despite his technical capacity, man operates under severe limitations, both in his ability to perceive and understand natural systems, and in his economic freedom of choice among different courses of action. It is to a large extent economic and social constraints which have, for example, led to the growth of some 2,000 towns of more than 1,000 people each on floodplains in the United States, and to the construction of some 125,000 structures (equivalent to the housing stock of Boston) less than 10 feet above mean sea level on the Eastern Seaboard between Maine and North Carolina. Table 8.2 gives a comparison of Great

TABLE 8.2

COMPARISON OF FARMERS' ESTIMATE OF DROUGHT
FREQUENCY WITH ACTUAL DROUGHT CONDITIONS IN THE PAST

	County	Adams	Barber	Frontier	Finney	Cimarron	Kiowa
	Farmers' estimate of drought (yrs./100)	*17*	*16*	*19·9*	*18·6*	*34·8*	*34·9*
Based on the Palmer Index	% Time drought	42·4	46·9	41·6	47·2	48·7	47·2
	% Mild drought and more severe	32·8	39·6	32·0	37·0	39·8	34·8
	% Moderate drought and more severe	23·6	26·8	20·8	26·6	30·8	24·4
	% Severe and extreme drought	15·7	13·8	11·2	15·4	18·4	13·4

(From Saarinen, 1966)

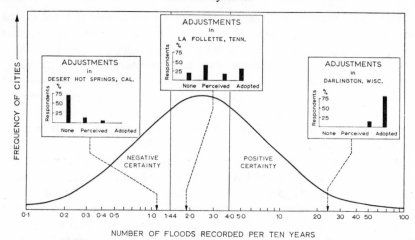

Fig. 8.4　　Logarithmic-normal frequency distribution of 496 urban places in the United States for which flood-frequency data are available. Examples of perception adjustment are given for three of these cities (Desert Hot Springs, California; La Follette, Tennessee; Darlington, Wisconsin). The adjustments exhibited by the respondents in these cities are scaled at four levels, varying from total ignorance (*none*), through two levels of perception (*perceived*), to adoption (*adopted*). Variations in these levels between the three cities are related to flood frequency (and even more so to *perceived frequency*), which is shown hypothetically as positive and negative certainty. (After Burton, Kates and White, 1968)

Plains farmers' estimates of drought frequency and actual drought conditions in the past. This brings out two important points; firstly, that men occupying areas of intermediate hazard occurrence are inherently optimistic; and, secondly, that in the more humid areas (Adams and Barber Counties), estimates of drought frequency correlate best with extreme drought conditions, whereas in the drier areas (Frontier, Finney and Cimarron Counties) they correlate best with moderate and more severe droughts. This indicates the important conclusion that the less probable the hazard, the greater the under-estimation of its danger. This matter is closely related to the kinds of magnitude and frequency analyses described in Chapter 5, in that the greatest environmental adjustment problems arise where there is an intermediate frequency of hazard, randomly distributed in time and space, combined with a high variability in the degree to which the members of the human population perceive the frequency of the hazard. Fig. 8.4 shows, for example, the different degrees of flood adjustments (in terms of four levels, varying from none, through two levels of perception, to adoption of measures) for three towns situated in regions of differing flood probabilities. In contrast, however, it can be shown that the inhabitants of similar sites often respond differently to environmental hazards. For example, the inhabitants of Rapid City,

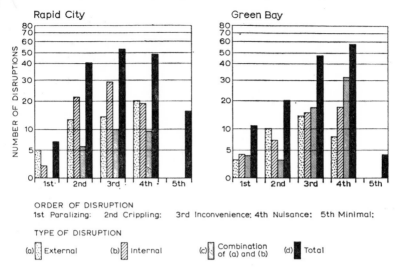

ORDER OF DISRUPTION
1st Paralizing: 2nd Crippling: 3rd Inconvenience: 4th Nuisance: 5th Minimal:

TYPE OF DISRUPTION

(a) External   (b) Internal   (c) Combination of (a) and (b)   (d) Total

Fig. 8.5   Comparison of five orders of snow disruption, each divided into four types, between Rapid City, South Dakota, and Green Bay, Wisconsin, for the years 1953–1963. (After Rooney, 1967)

South Dakota (mean annual snowfall 37 inches falling on 11·2 days) had only a marginal snow control programme geared to operate at the cessation of a blizzard, whereas Green Bay, Wisconsin (39 inches on 10·8 days) had an above average programme which was operated during blizzard conditions. Fig. 8.5 compares the number of 4 types and 5 orders of snow disruption in the two cities (1953–63), showing that Green Bay suffers less than 80% of the important snow disruptions which occur in Rapid City.

Human adjustment to natural hazards are, in some ways, typical of the ways in which man tries to come to terms with his environment. He can either attempt actively to manipulate natural systems in some manner (the main concern of this Chapter) or he can operate in a more passive manner by rearranging his behaviour within his socio-economic system. Fig. 8.6 depicts Kates' view of this two-fold approach to human adjustment in respect of hazards in which the *human use system* (defined as the smallest managerial unit capable of independent adjustment to the hazard) is compared with the *natural event system* (i.e. the magnitude, duration, frequency and temporal spacing of hazard-producing natural system states) to define a *natural hazard* to produce *hazard effects* and trigger off an *adjustment process control* subsystem leading to adjustments to the hazard involving *natural event modification adjustments* and/or *human use modification adjustments*.

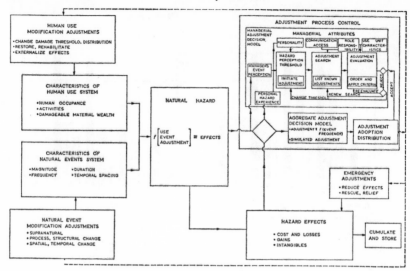

Fig. 8.6   A systems model of human adjustment to natural hazards .(After Kates, 1970).
A description is given in the text

8.3 Atmospheric Control Systems

Even a cursory survey of man's relationships with his immediate atmospheric
environment serves to emphasize two further important aspects of human
intervention in natural spatial process-response systems—the importance of
energy scales in determining the efficacy of man's actions, and the occurrence
of secondary unplanned effects. These aspects emerge with special clarity
in this context, firstly, because of the wide variety of spatial and energy
scales manifested by atmospheric phenomena and, secondly, because of the
relatively short relaxation times exhibited by this medium.

A. MICROSCALE

The construction of dwellings, apart perhaps from the adoption of clothing,
represents man's most traditional means of controlling his most immediate
atmospheric environment. The growth of large cities, particularly after the
Industrial Revolution, although these are micro features in terms of world
scale, has introduced atmospheric modifications both of a primary (planned)
and of a secondary (unplanned) character which are increasingly affecting the
lives of a larger and larger proportion of the world's population.

Large urban complexes exert considerable modifications over their local
atmospheric systems: by the addition of heat; by changes in the atmospheric

composition and heat balance; and by changes in such surface characteristics as roughness, albedo and hydrology.

Although more effective technological means of insulation and fuel utilization are tending to cut down urban heat loss to the atmosphere, cities are noticeably warmer than their immediate rural environs. In the mid-1950s, for example, the city of Hamburg was contributing some 40 langleys/day (i.e. 40 cal/cm²/day) to its atmosphere in December, compared with an amount of 34 langleys/day received as radiation from the sun and sky (i.e. $Q + q$; see Chapter 3). The 'heat island' effect of cities, however, is especially well marked at night with low wind speeds, and it seems clear that much of this thermal effect must be due to the absorptive and radiative properties of building materials.

Pollution of the urban atmosphere by gases such as sulphur dioxide and carbon monoxide and by solid particles (*aerosols*) is having far-reaching effects on the local solar energy cascade in such cities as Los Angeles and Tokyo. Fig. 8.7*a* shows the areal variation in received direct plus diffuse solar radiation ($Q + q$) over Downtown Los Angeles between 1400 and 1515 hours on 24th May 1968 (a very clear day), emphasizing the differences between the northern Central Business District and the southern residential area, on the one hand, and the middle industrial and commercial zone, on the other. The picture of net radiation balance ($R = (Q + q)(1 - \alpha) - I$; where $\alpha =$ albedo and $I =$ effective outgoing radiation) shows even more striking areal variation (Fig. 8.7*b*) with the effects of the tall vertical faces of the buildings in the Central Business District being distinguished from those of the relatively high proportion of vegetation in the residential area.

The net radiation balance thus involves the third microclimatic modification imposed by urban areas, the changes in surface characteristics. These changes include:

(*a*) The increase in surface roughness. This causes many phenomena, including convergence (i.e. piling-up) of moving air over cities, the tendency for tornadoes to 'lift off' the ground on encountering the rough urban surface, and highly variable wind speeds and directions.

(*b*) Change in the surface albedo, conductivity and heat storage capacity due to the character of the building and surfacing materials, and to the incidence of vertical faces of buildings. It has been estimated that European and American cities experience, on average, 11–18% more days with light rain and thunder than their surrounding countrysides.

(*c*) Alteration of the local hydrological characteristics by destruction of vegetation, elimination of standing water, and efficient surface drainage.

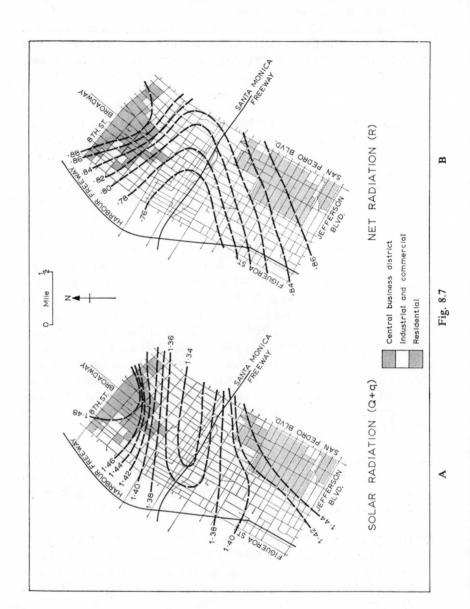

SOLAR RADIATION (Q+q)   NET RADIATION (R)

A   B

Central business district
Industrial and commercial
Residential

Fig. 8.7

B. MESOSCALE

The largest scale at which man can consciously intervene in atmospheric systems depends to a large extent upon the amounts of energy involved in their operation (Table 8.3), and it is very significant that more effective

TABLE 8.3

Total energy of various·individual phenomena and localized processes in the atmosphere (*from Sellers*, 1965)
[*Rates are relative to total solar energy intercepted by the earth* $(3·67 \times 10^{21} \, cal/day)$]

Solar energy received per day	1
Melting of average winter snow during the spring season	10^{-1}
Monsoon circulation	10^{-2}
World use of energy in 1950	10^{-2}
Strong earthquake	10^{-2}
Average depression	10^{-3}
Average hurricane	10^{-4}
Krakatoa explosion of August 1883	10^{-5}
Detonation of 'thermonuclear weapon' in April 1954	10^{-5}
Kinetic energy of the general circulation	10^{-5}
Average squall-line	10^{-6}
Average magnetic storm	10^{-7}
Average summer thunderstorm	10^{-8}
Detonation of Nagasaki bomb in August 1945	10^{-8}
Average earthquake	10^{-8}
Burning of 7000 tons of coal	10^{-8}
Daily output of Hoover Dam	10^{-8}
Moderate rain (10 mm over Washington, D.C.)	10^{-8}
Average forest fire in the United States, 1952–3	10^{-9}
Average local shower	10^{-10}
Average tornado	10^{-11}
Street lighting on average night in New York City	10^{-11}
Average lightning stroke	10^{-13}
Average dust devil	10^{-15}
Individual gust near the earth's surface	10^{-17}
Meteorite	10^{-18}

intervention has proved possible in local convective systems than in larger scale synoptic systems.

The most fruitful point of intervention in mesoscale atmospheric systems has been in the precipitation cascade where cloud seeding has been used to

Fig. 8.7   Radiation in Downtown Los Angeles between 1400 and 1515 hours on 24 May, 1968. (After Terjung, 1970)
(*a*) Direct plus diffuse solar radiation ($Q + q$) in langleys per minute.
(*b*) The net radiation balance ($R = (Q + q)(1 - \alpha) - I$).

stimulate condensation. Rain-making experiments of this type are based on three main assumptions:

1 Either the presence of ice crystals in a super-cooled cloud is necessary to release snow and rain (according to the Bergeron theory); or the presence of comparatively large water droplets is necessary to initiate the coalescence process.
2 Some clouds precipitate inefficiently or not at all, because these components are naturally deficient.
3 The deficiency can be remedied by seeding the clouds artificially with either solid carbon dioxide (dry ice) or silver iodide to produce ice crystals, or by introducing water droplets or large hygroscopic nuclei.

Such seeding is thus only productive under limited conditions of orographic lift and in thunderstorm cells, when nuclei are insufficient to generate rain by natural means. Natural precipitation occurs preferentially within certain upper-air temperature ranges—for example, some 80% of winter precipitation in the state of Washington falls when the 700-mb (10,000 ft.) temperature is not lower than $-10°C$ and is especially prevalent at $-4°C$ to $-8°C$. Artificial precipitation stimulation must exploit these preferences, and seeding is thus effective within a limited temperature range. Below $-20°C$. natural nuclei, such as dust, become active to form snowflakes, usually in sufficient numbers so that additional silver iodide particles are not needed, and under some conditions are actually detrimental. Cloud seeding may be effected by burning silver-iodide-impregnated fuels or solutions at ground level to produce a smoke which is carried upwards by wind into the effective zone, by firing rockets containing nuclei into the effective zone, or, more usually, by dropping the nuclei from aircraft.

Cloud seeding by these means has been attempted in many parts of the world, notably in Australia and the western United States. The need for fresh sources of water now and in the future is so acute in the United States that the Bureau of Reclamation has initiated the nation-wide Project Skywater to investigate the possibilities of water management through artificial rain-making. The purpose of one such scheme is to increase winter precipitation over the mountains of the Upper Colorado River, thus augmenting spring runoff, which would be stored and regulated to meet demand by the existing reservoirs. An increase of the November-April precipitation by 15% over 14,200 square miles of target areas is expected to yield an average additional runoff of 1,870,000 acre-feet annually. An important advantage of water provision by cloud seeding methods is that a 10% increase in rainfall can result in a 17–20% increase in runoff, because evaporation does not increase in proportion to the greater precipitation, so that there is more water available

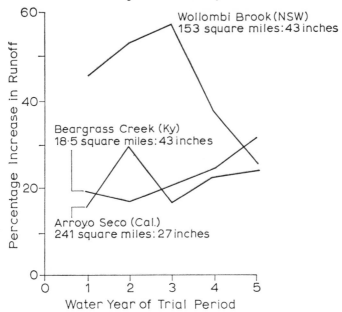

Fig. 8.8   A computer-based estimate of 5-year percentage runoff increases from three drainage basins in Australia, Kentucky and California resulting from a 10% increase in average precipitation. (After Crawford. In Sewell, 1966). Areas of the basins are given (in square miles), together with mean annual precipitation (in inches)

for runoff. Fig. 8.8 shows a computer-based estimate of 5-year runoff increases from three drainage basins, assuming an average precipitation increase of 10%. Preliminary investigations have shown that the optimum conditions for seeding are when there is a thin (less than 5,000 ft. thick) saturated air-mass layer, the temperature at the top of which is not less than −20°C, and the temperature over the target area is warmer than −10°C (Fig. 8.9). Eight major areas, lying generally above 9,500 ft. where annual natural runoff is over 10 in., contribute 75% of the total Upper Colorado basin runoff, although they form only 13% of the basin land area. Most precipitation comes in a few big storms, and since these storms are the main precipitation generators, it is important to take advantage of the limited opportunities they offer. Increasing the total precipitation, however, also increases the variability of its occurrence, since the fall from large storms is increased, the smaller rainfalls remaining the same. The cost of new water in the Upper Colorado, provided by cloud seeding, has been estimated to be approximately $1·00–$1·50 per acre-foot. (This is the operating and running cost, exclusive of the research necessary to make the work feasible.)

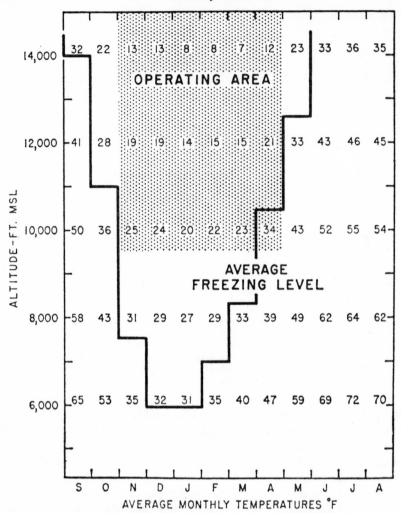

Fig. 8.9   An average temperature-altitude chart for Grand Junction, Colorado, showing the average freezing level and the optimum operating area for cloud seeding (i.e. between November and April above 9,500 feet). (After Hurley. From Chorley and More, 1969)

An associated method of artificially increasing precipitation is a corollary of cloud seeding and involves attempts to reduce the water loss by the evaporation of precipitation which takes place between the cloud base and the ground. In the Sonora Desert, Mexico, it has been estimated that, whereas 40 in. of rain is annually available at the cloud base, only 9 in. reaches the ground. The problem here is to find an agent which will increase the drop size and keep the drops large, and so far such an agent has not been found.

In summary, it is clear that there is a limited range of natural conditions in which significant artificial interventions can be made to produce, increase or conserve precipitation. The seeding of some cumulus clouds at temperatures of about $-10°C$ to $-15°C$ probably produces a mean increase of precipitation of some 10–20% from clouds which are already precipitating or 'are about to precipitate', with comparable increases up to 250 km. downwind, and increases of up to 10–15% have resulted from the seeding of winter orographic storms. On the other hand, the seeding of depressions has produced no apparent increases, and it appears that clouds with an abundance of natural nuclei, or with above-freezing temperatures throughout, are not susceptible to rain-making. At present it is often a difficult statistical matter to determine whether many of man's attempts have produced significant increases in precipitation; for example, six experiments in Washington and Oregon produced the following probabilities that rainfall had been increased: 95, 67, 50, 50 and 41%! Another instance serves also to highlight the possible legal problems which attempts at rain-making will provoke. In Quebec a recent 25% increase in rainfall coincided with rain-making attempts, causing extensive floods, crop damage, and disruption of the tourist industry. Following a large public outcry, the Federal Government announced that the effect of the seeding had been to decrease the possible rainfall receipt by 5%!

Besides rain-making, other human interventions in the mesoscale precipitation cascade involve the successful local dissipation of freezing fogs over airports by spraying with propane gas, brine or dry ice, causing snow to fall and clear the air. The Russians have also claimed success in dissipating damaging hailstorms by the use of radar-directed artillery shells and rockets to inject silver iodide into high-liquid-content portions of clouds, which freezes the available super-cooled water and prevents it from accreting as shells on growing ice crystals.

C. MACROSCALE

It is probable that man is on the brink of much larger interventions into the hydrological cycle on a scale of hundreds of square miles or more, although attempts to produce rain in synoptic-scale frontal systems have so far failed. There is, however, much evidence that the dramatic temperature increases observed over much of the globe during the past 100 years (especially in the winter temperatures of the northern mid-latitudes) may be partly attributable to the increase of CO_2 in the atmosphere due to the burning of fossil fuels by man.

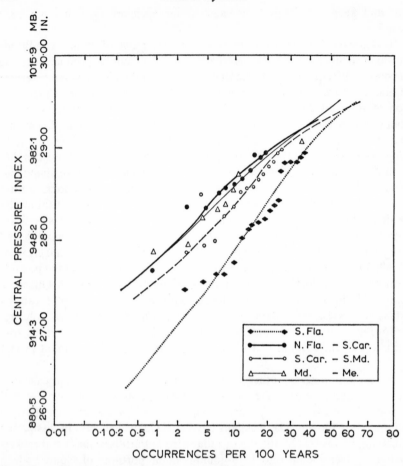

Fig. 8.10 Cumulative frequencies of the central pressure indices of hurricanes crossing the Atlantic coast of the United States (1900–1956), plotted as frequency per hundred years for each of four coastal sectors. (After Burton and Kates, 1964)

Hurricanes cause, on average, some $300 million worth of damage annually in the United States and Canada, and Fig. 8.10 shows the frequencies of hurricane centres crossing the Atlantic coast of the United States. At present the $9 million spent on forecasting, warning and protection is estimated to save some $25 million of property, with only about 20% of the affected population being involved in protective action. It has been estimated that improvements in forecasting and warning systems might increase the saving to some $100 million. More ambitious projects make it likely, however, that hurricanes can be suppressed or 'damped down' by the seeding of the rising air in the cumulus eye-wall, widening the ring of condensation and up-draught, decreasing the angular momentum of the storm and thus the maximum speed of its winds. The spreading of the sea ahead of the storm with oily materials might be used to cut off surface evaporation and thus the hurricane energy supply. It has been estimated that even if only modest reductions in storm intensity or slight changes in path could be achieved, U.S. annual hurricane losses might be cut by as much as one-third. Even such apparently beneficial attempts may represent potentially dangerous tampering with the natural global moisture economy, especially so in this instance when it is remembered that 30% of the August rainfall of the Texas coast, 30% of the September rainfall of the Louisiana and Connecticut coasts, and fully 40% of the September rainfall at Atlantic City, New Jersey, are derived from hurricane circulations.

Even more speculative schemes involve putting huge quantities of dust or metallic needles into stationary orbit to reduce sea temperatures locally and decrease evaporation; as well as creating 'thermal mountains' by painting desert surfaces black to increase their conservation of solar heat, stimulate convection, and thereby increase cloudiness and precipitations downwind. The unknown dangers attendant upon such large-scale tampering with the delicately-balanced earth-atmospheric system must postpone such schemes until theoretical mathematical models simulating its behaviour have been developed, so that all the possible effects can be predicted in advance by computers.

8.4 Terrestrial Control Systems

A. THE BASIN HYDROLOGICAL SYSTEM

No spatial process-response system has proved so susceptible to human control as the basin hydrological system. The relatively rapid and accessible movement of water on and near the earth's surface has long made this system liable to manipulation by man, beginning perhaps with primitive irrigation. Removal of vegetation is a common means of such intervention. It has been observed, for example, that the removal of the Douglas Fir cover from a small drainage basin in the Colorado Rockies increased the average streamflow run-off by 17% during the following three years, and raised peak flood discharges by 50%. Conversely, reforestation of an area of 88 acres in the Tennessee Valley decreased local peak discharges by up to 90%. More sophisticated chemical means of intervening in the vegetational part of the hydrological system are being developed as the result of work designed to decrease reservoir evaporation. The spraying of a monomolecular film of cetyl alcohol reduced evaporation from a 130 acre Tanzanian reservoir by some 12% (although working best for reservoirs of less than 1 acre, where the reduction is up to 25%), and maize sprayed with atrozine exhibited evapotranspiration reduction by

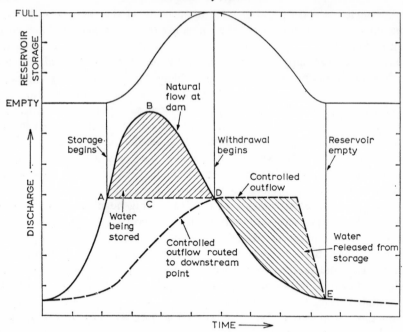

Fig. 8.11   The ideal flood-control operation of a reservoir (After Linsley, Kohler and Paulhus, 1949). A more complete description is given in the text

40–45% some 6 hours afterwards (the results for soya beans yielded a 60–70% reduction). Indeed, there is no field of human intervention in natural systems which is producing greater result than the hydrological system of individual drainage basins, ranging from traditional channel rectification to inhibit flooding to the treatment of snow surfaces with material of low albedo to stimulate melting.

However, still the most effective and economically viable means of human intervention in the basin hydrological system is through the construction of dams to control the streamflow cascade, although the purposes which these structures now serve have been multiplied and integrated. Chapter 8.6 gives examples of multiple-use planning, and for the present we will confine our remarks to the relatively simple case of flood-control dams.

It is clear that, if a reservoir can be constructed and maintained at the lowest possible level, it can provide a receptacle for flood waters which will have the effect of reducing a flood peak. Fig. 8.11 shows the ideal simple flood control operation of a reservoir. As the inflow hydrograph rises all inflow is discharged until point *A*, after which there is regulated discharge at a constant rate with excess inflow stored. When the inflow decreases to

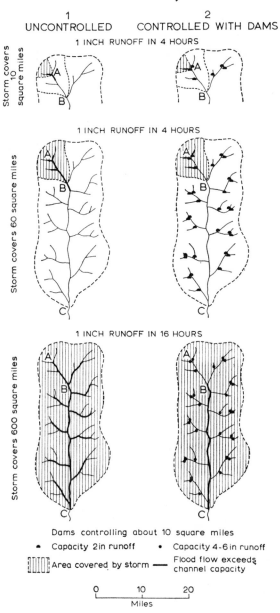

Fig. 8.12   The control exerted by dam construction on the flooding of a 600-square-mile drainage basin in respect of 2 different runoff intensities and 3 different storm sizes. (From *The Flood Control Controversy* by Luna B. Leopold and Thomas Maddock Jr. The Ronald Press Company, New York, 1954)

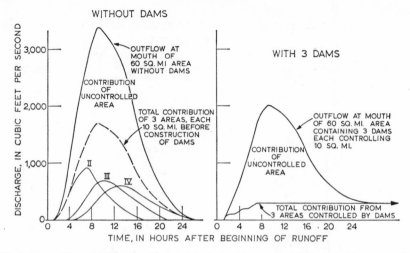

Fig. 8.13   The control exerted on a 1 inch in 4 hours runoff from a 60-square-mile second-order basin (*left*) by the constitution of 3 dams at the mouths of the first-order basins (each of 10 square miles in area) capable of temporarily retaining 2 inches of runoff (*right*). (From *The Flood Control Controversy* by Luna B. Leopold and Thomas Maddock Jr. The Ronald Press Company, New York, 1964)

this rate (*D*), the reservoir (in this ideal case!) is filled and the controlled discharge rate will be continued until the reservoir is again empty (*E*). The effect of this operation is to reduce the downstream flood peak from *B* to *C*. Obviously the size, number and location of dams exert important detailed controls over basin flooding. Fig. 8.12 illustrates these with respect to a 600 sq. mile third-order basin, containing 10 second- and 30 first-order basins of 60 and 10 sq. miles, respectively. Rainfall producing a runoff of 1 inch in 4 hours over a second-order basin (Fig. 8.12*b*) would, if uncontrolled, produce a runoff hydrograph at point *B* as shown in Fig. 8.13 (left). This hydrograph is composed of the respective runoffs from the three component first-order basins (II, III, IV) together with the uncontrolled interbasin area (30 sq. miles). The peaks of runoff for these three first-order basins reach point *B* after different time lags and combine with the contribution of the uncontrolled interbasin area to produce a hydrograph peak of 3,400 cubic feet per second some 9 hours after the beginning of runoff. The construction of three dams capable of temporarily retaining 2 inches of runoff (i.e. 10 cu. feet/sec./sq. mile) at the mouths of the first-order basins would produce the modification shown in Fig. 8.13 (right) with each dam releasing 100 cu. feet/sec. The hydrograph peak would thus be reduced to 2,000 cu. feet/sec. Fig. 8.12 compares the bank overflow in the basin resulting from three storm sizes (10, 60 and 600 sq. miles) and two intensities of storm

(1 inch of runoff in 4 hours and 1 inch in 16 hours), under controlled conditions and with dams at the mouths of the first-order basins of capacity 2 inches (Fig. 8.12a and b) and 4–6 inches of runoff (Fig. 8.12c).

B. THE EROSIONAL DRAINAGE BASIN SYSTEM

Man has become increasingly aware that general erosional rates in small drainage basins depend very much upon the intimate relationship between precipitation (amount and intensity) and vegetational cover. The effect of vegetation on relative erosion rates is shown in Fig. 8.14, indicating that, by weight, grass is much more effective than trees in retarding surface erosion, and Fig. 8.15 gives sediment yield for United States' drainage basins of between 10 and 50 square miles estimated from sedimentation rates in reservoirs. It is interesting that, under natural conditions, the maximum erosional rate occurs at an effective mean annual precipitation (i.e. the amount of precipitation required to produce the known amount of runoff) of about 12 inches where precipitation intensities are sufficiently great, and vegetational covers sparse enough, to combine to allow considerable removal of surface material by runoff. Obviously these relationships can be most effectively upset by human manipulation of the vegetational cover.

Strahler has developed a general theory for the steady-state developments of drainage basins in terms of:

1 A *geometry number* (HD/θ), where H = maximum basin relief;

D = drainage density and

θ = ground surface or channel gradient.

2 A *Horton number* (Qk_e), where Q = runoff intensity; and

k_e = an erosion proportionality factor (mass rate of debris removal per unit area/eroding stress per unit area).

This theory can be expressed as a dimensionless group:

$$\phi[(HD)/\theta, Qk_e] = 0$$

(where ϕ indicates non-dimensional, or dimensionless, quantities) showing that, for a given intensity of erosional process (i.e. Horton number) values of local relief, slope and drainage density reach a time-independent state in which the surface morphometry is adjusted to transmit through the system the quantity of debris and excess water characteristically produced

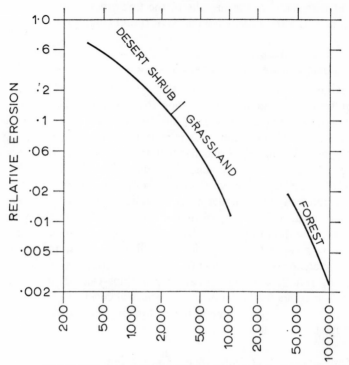

Fig. 8.14

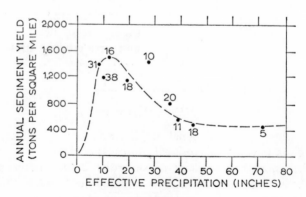

Fig. 8.15

under the controlling regime of climate. Man can upset this steady state by, for example, drastically decreasing the vegetational cover, causing the Horton number to increase and, in turn:

1 An increase of drainage density by gullying.
2 A moderate increase of θ by the steepening of stream channel and valley-side gradients (largely due to the generation of new low order channels with steep gradients and steep valley-side slopes).
3 A decrease in basin relief (measured as differences in elevation of head and mouth of first-order basins) due to headward erosion and valley-floor deposition. This channel aggradation is necessary to steepen gradients in order to remove the increased debris supply.

When this transformation is complete (Fig. 8.16), a new steady state is achieved by having a new set of landforms on a smaller scale than the original, but geared to a more intensive removal of debris.

C. DEBRIS CASCADES

Within broader spatial process-response systems geared to the transmission of water and debris, intervention is often very effective in terms of the inhibition of debris movement in linear cascades such as stream channels, valley-side slope profiles and shorelines. This often takes place inadvertently, as for example when the trapping of sediment in a reservoir causes the river to trench its channel downstream of the dam. This effect is shown in Fig. 8.17 in respect of the Elephant Butte and other dams on the Rio Grande. It is interesting that this local degradation has gone on during a time of general aggradation of the Rio Grande (see Fig. 7.7*b* for a location 40 miles upstream of Elephant Butte Dam), and that the latter tendency is noticeable at the lower part of the profile in Fig. 8.17. The employment of engineering works to stabilize natural slopes represents a very important area of human intervention in debris cascades, and the methods of effecting this are legion, including terracing and the decrease of pore-water pressure by artificial drainage.

Fig. 8.14   The effect of vegetation density on relative surface erosion rates in the United States. (From Langbein and Schumm, 1958)

Fig. 8.15   The relationship between the effective mean annual precipitation (i.e. the amount of precipitation required to produce the known amount of runoff) and annual sediment yields (estimated from reservoir sedimentation rates) for United States' drainage basins of between 10 and 50 square miles. The figures indicate the number of basins of given mean annual precipitation averaged to produce each point. (After Langbein and Schumm, 1958)

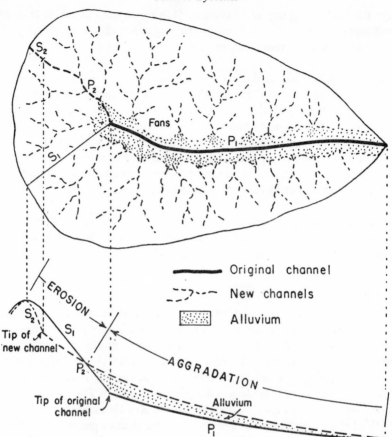

Fig. 8.16　A schematic impression of the transformation produced in an erosiona drainage basin due to a drastic decrease in the vegetational cover (From Strahler, 1958). A full description is given in the text.

Another common intervention in linear debris cascades is by means of groins to impede longshore beach drifting. This drift largely takes place within or just shoreward of the breaker zone as debris particles move both from the shoreward and seaward sides into the breaker zone within which they tend to follow elliptical longshore paths during the collapse of each wave. Fig. 8.18 compares the movement of fluorescent grains introduced at Santa Monica Beach (*A*) with that one mile north at Will Rogers State Beach (*B*), California, under rather similar breaker height and alignment. The effect of the groin on grain movement and on the resulting beach profiles at Will Rogers State Beach is immediately apparent. It should be pointed out, however, that the

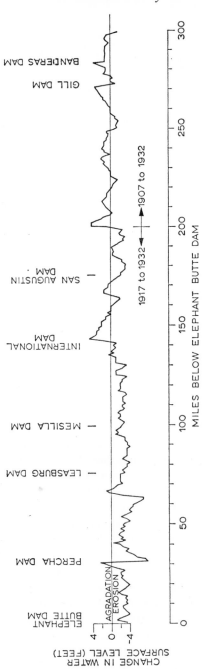

Fig. 8.17 Changes in surface elevation of the Rio Grande river downstream of Elephant Butte Dam, observed between 1917 and 1932 for less than 200 miles below the Dam; and between 1907 and 1932 for more than 200 miles below the Dam. (After Stevens, 1938)

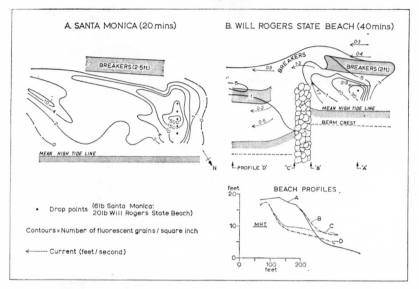

Fig. 8.18   The movement of fluorescent sand grains introduced inshore of the breaker zone at (*a*) Santa Monica Beach, California, and (*b*) near a groin at Will Rogers State Beach, California. (After Ingle, 1966)

pile-up of beach debris produces an erosional effect down the coastline analogous to the scour downstream of a debris-retaining dam.

D. GROUND-WATER SYSTEMS

A further group of examples of spatial physical process-response systems in which man is increasingly intervening is that which comprises the underground circulation of water. Increasing industrial, domestic and agricultural water demands are leading to sophisticated attempts to 'manage' ground-water resources by such techniques as artificial recharging of aquifers by water spreading or by direct pumping of water into the ground at times of excess, from whence it can be efficiently withdrawn during periods of deficiency. It is clear that the key valves in such a system are those of infiltration, seepage to the water table and discharge into stream channels.

An interesting example of a computer-based programme of ground-water control is provided by a scheme in West Pakistan where traditional methods of unlined canal irrigation had raised the water table to within 5 feet of the surface (Fig. 8.19*a*) leading to a capillary rise of salts which had formed surface deposits highly detrimental to agriculture (Fig. 8.19*b*). The basic problem was, firstly, to identify the key operational valves within the system

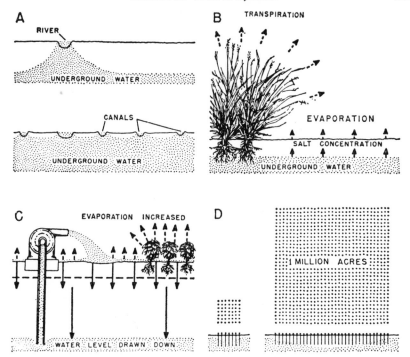

Fig. 8.19 Ground water relationships in irrigated areas of West Pakistan.

(*a*) The construction of leaky irrigation canals has allowed a general rise of the water table.

(*b*) Salt accumulates through the evaporation of saline water which rises into the topsoil under capillary action from the high water table.

(*c*) Waterlogging and soil salinity can be ameliorated by controlled cased tubewell pumping, which lowers the water table and supplies enough water at the surface to leach down soil salts before evaporation occurs.

(*d*) Large-scale tubewell pumping is required to negate the effects of lateral ground water seepage and to cause an appreciable lowering of the water table. (After Revell. From More, 1967)

(of which tubewell pumping of ground water was obviously a major one: Fig. 8.19*c* and *d*), and, secondly, to manage the system in such a manner as to reduce the level of the water table, improve drainage and cause the leaching out of surface salts during a period of continuing production of salt-resistant crops, the character and density of which regulated evapotranspiration to some extent. Fig. 8.20*a* gives a pictorial impression of the flow vectors involved in this problem, together with the key valves, and Fig. 8.20*b* depicts the same system as a flow diagram from which the computer

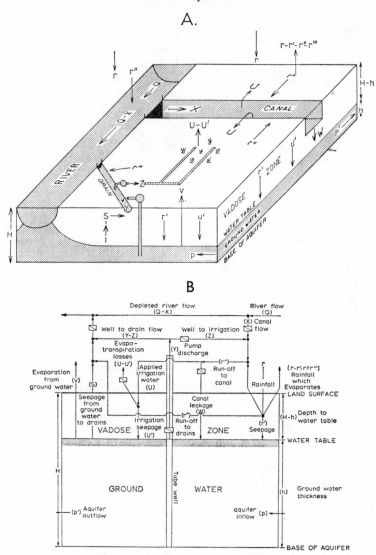

Fig. 8.20   A computer-based programme of control devices for combatting water-logging and salinity in the irrigated regions of West Pakistan, a full description of which is given in the text.

(*a*) A pictorial impression of the flow vectors involved in the problem.

(*b*) The same system depicted as a flow diagram from which the computer simulation programme was devised. The 'valves' are economic decision points—e.g. the setting of the valve at W represents the efficacy of the canal lining. (After Fiering. From More, 1967)

simulation programme was derived. The flow vectors were divided into three groups:

1 *Externally determined:* river flow (Q); rainfall infiltration to ground water (r'); overland flow to canals (r''); overland flow to drains (r''').
2 *Internally determined:* canal leakage (w:a function of lining); irrigation water applied (u); irrigation water infiltration (u'); ground-water seepage to drains (s); evaporation from ground water (v); local ground-water inflow (p); local ground-water outflow (p').
3 *Determined by operating policy:* canal flow (x); tubewell pumpage (y); tubewell water to irrigation (z); tubewell water to drain flow (y-z).

Within this system six valves were identified—w, x, y, z, s, and u-u' (the crop-controlled evapotranspiration). Relationships between the flow vectors were constructed into a computer simulation programme and 78 simulations involving likely combinations of variables were run. The optimum of these simulations, assuming a given u-u' and a tubewell spacing of 6,000 feet, was w = maximum (i.e. no canal lining), $s = 0$ (i.e. drains impermeably lined), $y = 1 \cdot 85$ ft^3/ft^2/year, and $z = 1 \cdot 85$ ft^3/ft^2/year.

8.5 Ecosystems

The position and relevance of biogeography within physical geography has traditionally presented problems, but the application of systems analysis effectively clarifies its role. At given levels, for example, the physical attributes of vegetation (e.g. amount of canopy cover) may constitute parameters of broader morphological systems or important storages in the solar energy or hydrological cascades. At higher levels of integration, ecosystems can form an effective spectrum of conceptual links between purely physical process-response systems and geographical control systems formed by the inter-penetration of physical and socio-economic process-response systems. At one end of this spectrum, some floral ecosystems are more akin to physical process-response systems dominated by mutual negative-feedback relationships, at the other the activity of the higher animals introduces an element of conscious intervention and positive-feedback into the systems operation so that this merges with that of the true geographical control systems manipulated by *homo sapiens* for purposes of resource management.

There are many possible definitions of an ecosystem, but perhaps the most useful for our purpose is that given by Lindeman in 1940. An ecosystem, in his view, is 'any system composed of physical—chemical—biological processes within a space-time unit of any magnitude'. From this definition it is clear that ecosystems, like all systems of interest to physical geography, may be defined at all scales of magnitude and complexity and that problems will arise in delimiting the boundaries of any one ecosystem. Clearly all ecosystems are process-response systems and equally clearly they are, in themselves, control systems since the living components—bacteria, plants and animals— act upon and modify not only the more inert physical components of the environment but control the flow of solar energy and the cycling of water and nutrients through the biosphere as a whole.

The crucial importance of natural ecosystems is, therefore, twofold. In themselves they act as control systems with respect to most features of the environment of this planet; they further represent a major point at which human control systems must intersect with the natural world.

Before we turn to a discussion of the latter subject, it is, perhaps, necessary to outline the manner in which a simple terrestrial ecosystem operates. Such a structure possesses the following characteristics; a location in space; con- tinuing inputs of solar energy; cycling of water and nutrients; autotrophic organisms (green plants) capable of photosynthesis and the conversion of a fraction of the incoming solar energy to chemical energy; heterotrophic organisms, deriving energy from the autotrophs (these may be of two main kinds: macro-consumers and decomposers). The major linkages between these components are shown diagrammatically in Fig. 8.21—this picture is not, of course, complete, as it omits, for the sake of clarity, the various direct links between the hydrological cascade and the autotrophs (see Chapter 3). If an ecosystem is in equilibrium, then it is clear that the output of chemical energy and heat (as plant and animal tissues, reflected radiation and respira- tion) must ultimately equal the input of solar energy; and that any loss of nutrients as a result of cycling between autotrophs and heterotrophs must be balanced by gains from the atmosphere and weathering of the bedrock. Obviously, neither of these conditions will be met exactly in any natural system which will not have fixed, closed, boundaries so that there will be an export of both chemical energy and nutrients as the living or dead bodies of plants and animals are moved into adjacent systems. One may, however, consider that such chance exports will be balanced, in the long-run, by equally random imports of individuals or debris, such that there will be no systematic gain or loss of mass and energy from any one ecosystem under natural conditions.

It has been suggested that there is a close resemblance between ecosystems and erosional systems, particularly in the humid tropics, as a result of the high proportion of solar energy involved in the vegetational processes. High vegetational productivity leads to a high rate of regeneration and this, in turn, may favour great depths of weathering. However, dense vegetation may not permit great erosional rates and it is often in a delicate balance with the soil, favouring not only its formation but its storage. Artificial removal of vegeta- tion under these conditions results in rapid removal of soil from this storage

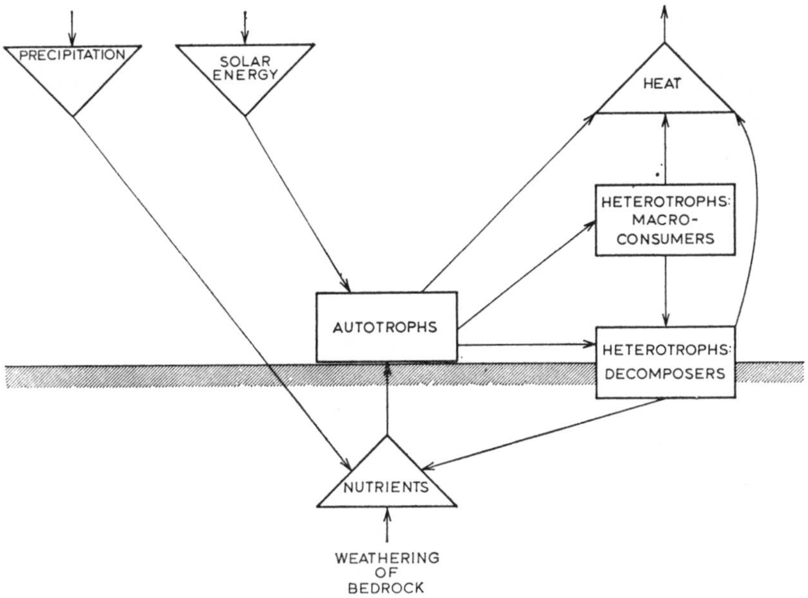

Fig. 8.21   The major linkages between the components of a simple terrestrial ecosystem.

and in the high rates of denudation which are locally observed in the humid tropics.

Perhaps the most important control of the operation of ecosystems is that exerted by the particular resource which is in shortest supply: this is known as *Liebig's Law of the Minimum*. Obviously in many environments, the solar energy receipt is an absolutely limiting factor (as, for example, in the deep oceans); in others (hot and cold deserts), water will act as the ultimate control of productivity. In most ecosystems, however, it is very difficult to isolate the limiting factor, which may be a macro- or micro-nutrient (trace element) or even an unfilled ecological niche. Under natural conditions an ecosystem will be adjusted to give constant productivity within the constraints imposed by the limiting factor. More important still, ecosystems—like all biological systems—exhibit *homeostasis:* that is, the ability to compensate for fluctuations in any part of their environment by negative feedback mechanisms. The more complex the ecosystem, the greater the homeostatic potential. It is, therefore, much more unlikely that catastrophic fluctuations in productivity will occur in a well-established tropical mainland ecosystem than in the environment of a newly-emerged arctic island. (Complexity of ecosystems increases, generally, with annual radiation total, age of ecosystem and size of the land area on which it is found.)

23   (24 pp.)

Perhaps the major result of human intervention in ecological control systems is the reduction of natural complexity. This has the immediate effect of reducing the homeostatic potential and consequently allowing limiting factors to operate with greater severity. The replacement of natural ecosystems by those under human control is obviously a very tricky performance. On the one hand it is necessary for many ecosystems to be converted from uneconomic, albeit high-level natural productivity to perhaps lower-level but economically beneficial production (not *all* artificial ecosystems are relatively unproductive, however: sugar-cane plantations, for example, are among the most efficient converters of solar to chemical energy). On the other hand, to ensure that a highly simplified system continues to be productive requires additional human intervention in order to create the necessary negative feedbacks; fertilization of cultivated fields depleted of nutrients by cropping is an obvious example. This artificial compensation for the lack of natural regulatory mechanisms requires additional inputs of energy into the system: and 'energy' here may be considered to have four forms: manpower; mechanical energy; chemical energy; capital. What is required is a sufficient understanding of the way in which the system operates, so that economic productivity is maximized while the energy inputs required to maintain the system in balance are minimised.

Different types of agricultural activity over the ages have approached this problem in different ways. The shifting cultivation that was characteristic of much primitive farming dealt with the difficulty by moving the site of cultivation every few years: here the prime additional inputs involve the time and energy required to transport the cultivators to a new and undisturbed site. Similar principles apply to nomadic pastoralists. In both cases there are large-scale diseconomies involved in the proportion of land lying uncropped at any one time, but representing the 'bank' of potential productivity. Settled agriculturists, stock- and tree-farmers have approached the basic problem in quite another way, in general. They have, effectively, argued that the increase in economic productivity and saving in labour, time and space to be gained from monocultures of particular species in particular areas outweighs the diseconomies of fertilizers, irrigation, rotation, fencing, and pest and disease control. All of these artificial protective or homeostatic mechanisms require the expenditure of energy in one form or another to maintain the artificial system in balance. What is now becoming apparent is that, in many cases, the highly artificial ecosystem of the monoculturalist in fact requires far too many of these additional energy inputs to be viable. There is, in fact, a vicious circle established and the further the ecosystem departs from natural complexity, the greater the economic cost—gauged on the broadest scale—of maintaining long-term productivity. For example, on average, the equivalent of 8% of the value of the U.S. cotton crop in

1963 was spent on defensive sprays, weedkillers and defoliants, while 10 %
of the value of the banana production of the Windward Islands was needed for
similar protective measures. It is the realization of the magnitude of these and
other costs that is causing agriculturalists and economists to look towards a
measure of increased diversity as an economic benefit. To give a very simple
example, it has been found in South Africa that yields from cabbages increase
if certain weeds are allowed to remain in the ecosystem. This is because the
weeds provide a year-round home both for a moth which attacks the cabbages
and for its own natural enemies: as both pest and its control are present all
year round, the former is kept within bounds and the production of cabbages
increased. Where clean-weeding is employed, the moth arrives at the cabbage
field well ahead of its predators and is able to expand its numbers unchecked,
unless artificial control measures are employed.

It must be stressed that most natural ecosystems are complex, highly
productive, but largely valueless to man directly. Artificial ecosystems are
simple, generally rather less productive, but of immense immediate value.
What is increasingly being realized is that, in order to assess the true benefit
of an ecosystem, man must be able to weigh not only the immediate cash
ratio of input/return, but the long-term benefits of the ecosystems as a
control over the environment (for example, the effect of natural forest in
limiting runoff and soil erosion) and the often almost intangible advantages
of complexity as a source of homeostatic mechanisms. As with most systems,
the components of ecosystems are so interlocked that man may consciously
or unconsciously operate a 'valve' within any one at almost any point.
This, in itself, as we have said, is economically necessary. What is being
realized is that we must retain, as far as possible, something approaching the
natural multiplicity of regulators. In other words, we must endeavour to
work *with* the ecosystem and not against it, in order to minimize the energy
required to both satisfy our immediate economic needs and retain at least
indirect control over the greater part of our environment.

8.6 The Socio-Economics of Control Decisions

It is important to recognize that, in the foregoing examples of human intervention into spatial process-response systems, not only has the full range of
actual intervention not been explored, but also the scale on which con-
ventional intervention is technically possible has been underestimated. For
example, sound engineering plans are in existence for the large-scale transfer
of water from Alaska to Mexico and from northern Russia to the steppes,
combined with huge power, irrigation and navigation projects. Whether
these and similar schemes are ever implemented depends primarily on socio-
economic decisions involving the quality of the environment and, chiefly,
their costs in relation to alternative projects. Decisions of this kind are made

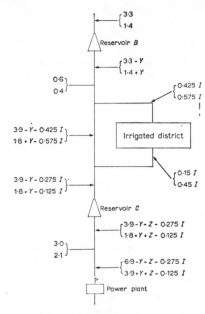

$\begin{cases} 3\cdot3 \\ 1\cdot4 \end{cases}$

Reservoir *B*

$\begin{cases} 3\cdot3 - Y \\ 1\cdot4 + Y \end{cases}$

$\left.\begin{matrix} 0\cdot6 \\ 0\cdot4 \end{matrix}\right\}$

$\begin{cases} 0\cdot425\,I \\ 0\cdot575\,I \end{cases}$

$\left.\begin{matrix} 3\cdot9 - Y - 0\cdot425\,I \\ 1\cdot8 + Y - 0\cdot575\,I \end{matrix}\right\}$

Irrigated district

$\left.\begin{matrix} 3\cdot9 - Y - 0\cdot275\,I \\ 1\cdot8 + Y - 0\cdot125\,I \end{matrix}\right\}$

$\begin{cases} 0\cdot15\,I \\ 0\cdot45\,I \end{cases}$

Reservoir *C*

$\begin{cases} 3\cdot9 - Y - Z - 0\cdot275\,I \\ 1\cdot8 + Y + Z - 0\cdot125\,I \end{cases}$

$\left.\begin{matrix} 3\cdot0 \\ 2\cdot1 \end{matrix}\right\}$

$\begin{cases} 6\cdot9 - Y - Z - 0\cdot275\,I \\ 3\cdot9 + Y + Z - 0\cdot125\,I \end{cases}$

Power plant

Fig. 8.22   A hypothetical stretch of river with mean wet- (*above*) and dry-season (*below*) discharges (10^6 acre-feet) shown in brackets for each reach. Possible sites for two dams, an irrigated district and a power plant are indicated. A full description is given in the text. (From Dorfman, 1962)

even more complex both by the long timescale over which they are implemented, and by the growing need for them to be involved in multiple-purpose projects the various parts of which have to be integrated together and with larger external social and economic objectives. The designs of water-resource systems, in particular, have been subjected to detailed economic analysis and the two examples which follow involve, respectively, a simple programming approach to a hypothetical problem and the application of systems analysis and computer simulation to a real-world situation.

Fig. 8.22 depicts a hypothetical stretch of river with mean wet and dry season discharges (in 10^6 acre feet) shown in brackets for three reaches. There are two possible sites (*B* and *C*) for reservoirs; an irrigated district; and the need for a downstream run-of-the-river power plant. The problem, described by Dorfman, is to determine the optimum magnitudes of four *decision variables:*

Y = the active capacity of reservoir *B* (this can be filled during the wet season and drawn on during the dry season to maintain downstream flow).

I = the annual amount of irrigation water required (42·5% being needed in the wet season; 57·5% in the dry season).

Z = the active capacity of reservoir *C* (operated similarly to *B*).

E = energy output of the power plant.

The above situation places a number of constraints on the solution for Y, I, Z and E:

The decision variables must obviously be non-negative:

$$Y \geqslant 0 \qquad (1)$$
$$I \geqslant 0 \qquad (2)$$
$$Z \geqslant 0 \qquad (3)$$
$$E \geqslant 0 \qquad (4)$$

Discharges in all reaches must be non-negative:
(if equations 5–8 are satisfied, all reaches are satisfied).

$$3 \cdot 3 - Y \geqslant 0 \qquad (5)$$
$$3 \cdot 9 - Y - 0 \cdot 425I \geqslant 0 \qquad (6)$$
$$1 \cdot 8 - Y - 0 \cdot 575I \geqslant 0 \qquad (7)$$
$$3 \cdot 9 - Y - Z - 0 \cdot 275I \geqslant 0 \qquad (8)$$

The following simple relationship exists:

(where: E = energy generated, in 10^9 *KWH* and F = flow through the turbines during the same period, in 10^6 acre feet).

$$E = 0 \cdot 144F \qquad (9)$$

If half the power must be generated in each part of the year:

Flow in wet season $= 6 \cdot 9 - Y - Z - 0 \cdot 275I \geqslant 0 \cdot 5E/0 \cdot 144 \, (= 3 \cdot 74E)$

Flow in dry season $= 3 \cdot 9 - Y - Z - 0 \cdot 125I \geqslant 0 \cdot 5E/0 \cdot 144 \, (= 3 \cdot 74E)$

Re-arranging:

$$Y + Z + 0 \cdot 275I + 3 \cdot 47E \leqslant 6 \cdot 9 \qquad (10)$$
$$- Y - Z + 0 \cdot 125I + 3 \cdot 47E \leqslant 3 \cdot 9 \qquad (11)$$

Decisions regarding the magnitudes of the decision variables are approached by setting up the following *objective function:*

$$\pi = B_1(E) + B_2(I) - K_1(Y) - K_2(Z) - K_3(E) - K_4(I)$$

where π = present value of net benefits (in 10^6 dollars).

$B_1(E)$ = present value of an output of $E \times 10^9$ *KWH* per year ($\$10^6$).

$B_2(I)$ = present value of an irrigation supply of $I \times 10^6$ acre-feet/year ($\$10^6$).

23 A

$K_1(Y)$ = capital cost of building reservoir B to capacity Y per year ($\$10^6$).

$K_2(Z)$ = capital cost of building reservoir C to capacity Z per year ($\$10^6$).

$K_3(E)$ = capital cost of building power plant to capacity E per year ($\$10^6$).

$K_4(I)$ = capital cost of building irrigation system to capacity I per year ($\$10^6$).

It is possible to solve the above simultaneous equations and obtain the functions with varying degrees of difficulty. For example, the first three capital-cost functions can be simply obtained from economic rule-of-thumb, where increasing scale brings increasing returns:

$$K_1(Y) = 43Y/(1 + 0\cdot2Y)$$

$$K_2(Z) = 47Z/(1 + 0\cdot3Z) \quad \text{(Fig. 8.23)}.$$

$$K_3(E) = 20\cdot6E - E^2$$

The non-linear relationships involved in other calculations present problems, but it is possible to produce a series of equations involving Y, I, Z and E. When these are solved for this simple case, the values of the decision variables are $Y = 0$ (i.e. no reservoir B needed), $Z = 1\cdot275$ (10^6 acre feet), $E = 1\cdot3834$ (10^9 KWH) and $I = 3$ (10^6 acre feet); producing a value of π of $\$494\cdot6$ ($\times10^6$).

This hypothetical and simplified example of the programming approach to the taking of economic decisions regarding intervention to develop water resource systems is illustrative of the kind of approach which was applied to environmental decision making more than a decade ago. The obvious difficulties of such an approach fall into two groups:

1 The objective function is usually excessively complex and can only be evaluated by a series of *iterations*, assumptions and approximations. This shortcoming is not as important, however, as:

2 The over-simplified approach which makes for a rigid and distinctly unrealistic approach to much more complex real-world situations. For example, the hydrological and economic environments are constantly changing, with discharges and interest rates susceptible to short-term fluctuations. The simple programming approach clearly is not sufficiently dynamic enough to take account of these variables.

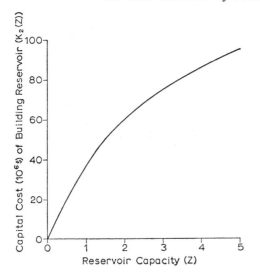

Fig. 8.23 The function relating reservoir capacity to capital building costs. (From Dorfman, 1962)

Much of the research into the economics of resource development which has been carried on during the 1960s has been concerned with the development of approaches which are much more geared to realistic changes of inputs and constraints, often containing a strong stochastic element, and to a greater flexibility in terms of changes of investment and of target output levels during a number of time periods within the history of the project, The most important of such approaches is represented by *systems analysis*.

Hufschmidt and Fiering have produced a systems-based computer model for the allocation of water resources in the Lehigh Valley, Pennsylvania. This begins with the assumption that three sets of phenomena must be interrelated, and these are designated as the supply configuration (Fig. 8·24a), water demands and problems (Fig. 8.24b), and possible development measures (Fig. 8.24c). The Lehigh problem primarily involved the construction of a number of reservoirs—the number, sizes and locations of which had to be determined—in order to provide:

1 Regulated river flows for water supply and water quality improvement at Bethlehem.
2 Recreation at the reservoirs.
3 Storage for flood protection.
4 Hydro-electric power production.

After preliminary examination it appeared that the possible development measures included a maximum of 6 reservoir sites (with sites 1 and 2 being

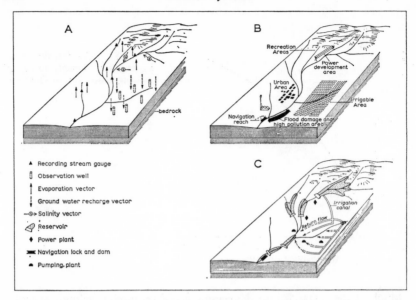

Fig. 8.24   The three sets of phenomena which must be integrated in producing a systems-based computer model for the allocation of water resources. (From Hufschmidt and Fiering, 1967)

(*a*) Supply configuration.
(*b*) Water demands and problems.
(*c*) Development measures.

alternatives), one of which (site 3) could be fed by a diversion channel. Reservoirs 3–6 could support 1 HEP plant each and reservoirs 1 and 2 more than 1 HEP plant apiece. In all, there seemed to be 5 theoretical combinations of HEP plants which were economically possible (Fig.8.25). The simulation model contained, therefore, 42 major design variables: 24 related to allocation of reservoir capacity (6 reservoir allocations of dead storage capacity, 6 of flood storage capacity, 6 of target reservoir levels for recreation, and 6 adjusted reservoir recreation levels); 6 related to the reservoir locations; 9 to possible power plant sites; 1 to the presence or absence of the diversion channel; 1 related to the target output of water supply; and 1 related to the target output of energy. The dynamism of the model was largely provided by the fact that 10 time changes of capital investment and target output levels could be accommodated, giving an effective total of 52 (i.e. 42 + 10) major design variables. The magnitude of the decision problem now begins to become apparent because, even if each of these variables could only assume one of three values (i.e. 'high', 'medium' or 'low'), many millions of combinations are possible!

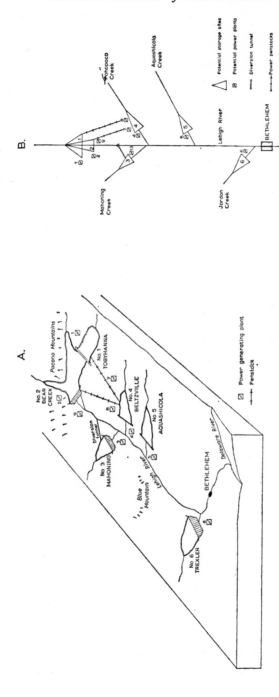

Fig. 8.25 The possible development measures for the Lehigh Basin water-resource simulation study. (From Hufschmidt and Fiering, 1967). (*a*) As a block diagram, and (*b*) schematically.

The mathematical model was designed to receive inputs of 12 mean monthly flows, together with 3-hourly flows for synthetic flood peaks, and mathematical relationships were developed governing the distribution of these flows through the system, on the basis of which the simulation was conducted. The object of the model was, assuming given system inputs, to simulate the various combinations of design possibilities on the assumption of the following order of social priority:

1 Stream flows and available reservoir flood storage capacities must be maintained above minimum amounts, as the top prority.
2 Reservoir levels to be kept up as far as possible to heights sufficient for recreation activities.
3 Given water supply outputs must be maintained as far as possible.
4 Given energy outputs must be maintained as far as possible, although this had lowest priority.

Given the above inputs, the simulation programme was required to explore combinations of the major design variables, operating according to the rules of priority, to suggest optimum combinations of variables for given lengths of simulation and planning demand periods.

Clearly, the total number of possible combinations had to be drastically reduced before simulation was a practical possibility, and, to this end, a series of *benefit/cost analyses* were conducted relating dam size costs to the benefits resulting from energy production, water supply, recreation facilities provided, and reductions in flood peaks in order to give practical limits within which the simulation of different combinations of design variables could take place. Even this reduction was not sufficient, however, and the model was eventually used to explore likely combinations of variables which were randomly selected to give a wide range of capital investment and running costs, on the one hand, and of benefits accruing from energy production, water supply, recreation and flood control, on the other. Of these sampled combinations, only 3 showed net benefits over a 50-year period:

1 Involving a large capital investment, 5 reservoirs (1 and 3–6), and high target levels of energy production and water supply.
2 A slight variant on 1.
3 Involving a relatively low capital investment, 3 reservoirs (3–5), and an extremely low energy production target.

These three schemes were used as the basis for further investigations and a final scheme was produced yielding a benefit/cost ratio of 1·8.

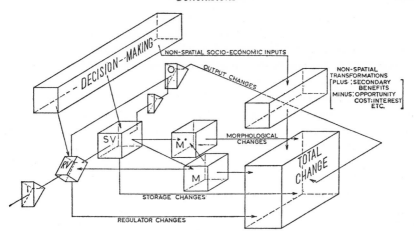

NON-SPATIAL SOCIO-ECONOMIC INPUTS

NON-SPATIAL
TRANSFORMATIONS
[PLUS : SECONDARY
BENEFITS
MINUS: OPPORTUNITY
COST:INTEREST
ETC.]

GEOGRAPHICAL CONTROL SYSTEM

Fig. 8.26   A geographical control system in which concerted decision making achieves spatial expression in effecting output, morphological, storage and regulator changes, as well as performing non-spatial transformations (see Figs. 1.3, 8.1 and 8.2). (From Chorley, 1971)

The foregoing description is indicative of the increasing complexity of decision-making involving environmental intervention on a large scale, and the reliance which is being placed on the use of large computers programmed on a systems basis.

8.7 Conclusion

In this chapter we have been able only to touch briefly on some of the ways in which dominantly non-spatial socio-economic systems intersect with dominantly spatial process-response systems to produce the stuff of what is coming to be termed 'physical geography' if one is primarily concerned with the latter, 'human geography' if one is mainly concerned with the spatial manifestations of the former, and 'geography' if the intersection problems preoccupy one. Fig. 8.26, attempts to illustrate something of the nature of this intersection which produces a *geographical control system*. Herein lies the heart of much of the current geographical dilemma; namely, that decision-making systems are extremely complex, are not inherently spatial in structure, and operate very largely in terms of the *cumulative causations* characteristic of positive feedback. Such systems are characterized by innovation, growth, elaboration and change, and, when linked to physical process-response systems which are essentially conservative but susceptible of being impelled through significant operational thresholds, present immense conceptual and practical problems to those who are concerned with interactions of the animate and inanimate worlds. This problem of interaction is continually being aggravated by the increasing subordination of much of the physical environment to the effects of human action, thus disrupting the 'man-land' duality which has formed the basis of much traditional geography.

However, although the importance of the physical parts of such geographical control systems may be quite small in comparison with associated socio-economic components, two important attributes contributed by the physical process-response systems are ingredients necessary to the creation of a truly geographical system. The first ingredient is some consideration of the systematic operation of spatial physical processes. Without this the whole system would rest purely upon social and economic variables, and it is very characteristic that the application of systems analysis in 'regional science' is virtually unconcerned with physical variables, which enter only as minor economic throughputs or constraints. The second ingredient contributed by physical process-response system components to the larger geographical systems is that to do with spatial location, spatial magnitude, and their inter-relationship. It is also characteristic of social and economic systems that their specialized locational and geographical components do not usually amount to much more than considering regions as individual 'boxes' subject to input/output analysis. It is a basic attribute of process-response systems that the subsystems of which they are composed are linked geographical units having both spatial magnitude and location. Thus it is only when one views the physical process-response systems as forming a part of the larger control system that the whole can gain both the physical ingredient, however subsidiary this may be in individual instances, and a true regional and locational flavour.

What is the implication of the foregoing thesis in terms of teaching and research in the discipline historically designated as 'physical geography'? It is, we suppose, that there can be no second-class citizens among those who, for whatever reason, study the nature and operation of spatial process-response systems involving surface terrestrial processes. Whether one wishes to treat them as control systems, forming parts of larger geographical systems, or to exploit their longer time-span operational relationships in terms of the transformations often more relevant to the earth sciences, is immaterial in so far as the basic necessity for understanding the structure of the process-response system is concerned. All students of earth-scale phenomena should, therefore, occupy themselves with these spatial process-response systems, although ultimately the earth scientist and the geographer may go their separate ways. The former will continue to deepen his understanding of physical process-response systems, particularly in terms of their time-bound development within a timescale which is often of a larger order of magnitude than that of human history. The latter will concentrate on the equilibrium relationships of the process-response systems and the manner in which they can either be disturbed by inadvertent human intervention leading to a degradation of earth resources, or thoughtfully regulated so as to exploit their inherent operational characteristics as a part of wider geographical systems being controlled for the well-being of a wide range of living things— man included. It is only through this application of systems analysis that considerations of the management of the natural environment can be elevated above mere *ad hoc* book-keeping to form part of a broader scholarly discipline which focuses on the conservational aspects of geographical control systems.

REFERENCES

Ackoff, R. L. (1960), Systems, organizations and interdisciplinery research; *General Systems Yearbook*, **5**, 1–8.

Ashby, W. R. (1964), *An Introduction to Cybernetics* (Methuen, London), 295.

Barry, R. G. and Chorley, R. J. (1971), *Atmosphere, Weather and Climate* (Methuen, London), 379.

Beer, S. (1966), *Decision and Control: The Meaning of operational research and management cybernetics* (Wiley, London) 556.

Bowden, L. M. (1965), Diffusion of the decision to irrigate; *University of Chicago, Department of Geography, Research Paper*, **97**, 146.

Bruce, J. P. and Clark, R. H. (1966), *Introduction to Hydrometeorology* (Pergamon, Oxford), 319.

Buckley, W. (1967), *Sociology and Modern Systems Theory* (Prentice-Hall, New Jersey), 227. ·

Burton, I. and Kates, R. W. (1964), The floodplain and the seashore; *Geographical Review*, **54**, 366–385.

Burton, I., Kates, R. W. and White, G. F. (1968), The human ecology of extreme geophysical events; *Natural Hazard Research, Working Paper*, **1**, 33.

Chorley, R. J. and More, R. J. (1969), The interaction of precipitation and man; In Chorley, R. J. (Ed.), *Water, Earth and Man* (Methuen, London), 157–166.

Chow, Ven Te (Ed.) (1964), *Handbook of Applied Hydrology* (McGraw-Hill, New York).

Cohen, S. B. and Rosenthal, L. D. (1971), A geographical model for political systems analysis; *Geographical Review*, **61**, 5–31.

Dorfman, R. (1962), Mathematical models: A multistructure approach; Ch. 13. In Maass, A. *et al, Design of Water Resource Systems* (Macmillan, London), 620.

Douglas, I. (1967), Natural and man-made erosion in the humid tropics of Australia, Malaysia and Singapore; "*Symposium on River Morphology*", *International Association of Scientific Hydrology, General Assembly of Bern*, 17–29.

Douglas, I. (1969), The efficiency of humid tropical denudation systems; *Transactions of the Institute of British Geographers*, No. 46, 1–16.

Eliot Hurst, M. E. (1968), *A Systems Analytic Approach to Economic Geography;* Association of American Geographers, Committee on College Geography, Publication No. 8.

Fiering, M. B. (1965), Revitalizing a fertile plain; *Water Resources Research*, **1**, 41–61.

Flawn, P. T. (1970), *Environmental Geology: Conservation, land-use planning and resource management* (Harper and Row, New York), 313.

Foote, D. C. and Greer-Wootton, B. (1966), Man-environment interactions in an Eskimo hunting system (Mimeo.); *Paper presented at Section H, Anthropology, 133rd Annual Meeting of the American Association for the Advancement of Science, Washington, D.C.*

Gates, G. R. and Anvari, M. (1970), A water balance control-theoretic model for regional management; *Geographical Analysis*, **2**, 19–29.

Gould, P. R. (1963), Man against his environment: A game theoretic framework; *Annals of the Association of American Geographers*, **53**, 290–297.

Hall, A. D. (1962), *A Methodology for Systems Engineering* (Van Nostrand, Princeton), 478.

Hall, W. A. and Dracup, J. A. (1970), *Water Resources Systems Engineering* (McGraw-Hill, New York), 372,

Hare, V. C. (1967), *Systems Analysis: A diagnostic approach* (Harcourt, Brace and World, Inc., New York), 544.

Harrison, C. M. and Warren, A. (1970), Conservation, stability and management; *Area*, No. 2, 27–32.

Hufschmidt, M. M. and Fiering, M. B. (1967), *Simulation Techniques for Design of Water-Resource Systems* (Macmillan, London), 212.

Ingle, J. C. (1966), The movement of beach sand; *Developments in Sedimentology*, **5**, (Elsevier, Amsterdam), 221.

Isard, W. *et al.* (1968), On the linkage of socio-economic and ecological systems; *Papers of the Regional Science Association*, **21**, 79–99.

Kates, R. W. (1962), Hazard and choice perception in flood plain management; *University of Chicago, Department of Geography, Research Paper* **78**, 157.

Kates, R. W. (1970), Natural hazard in human ecological perspective; Hypotheses and models; *Natural Hazard Research, Working Paper* **14**, 26.

Langbein, W. B. and Schumm, S. A. (1958), Yield of sediment in relation to mean annual precipitation; *Transactions of the American Geophysical Union*, **39**, 1076–1084.

Leopold, L. B. and Maddock, T. (1954), *The Flood Control Controversy* (The Ronald Press, New York), 278.

Linsley, R. K., Kohler, M. A. and Paulhus, J. L. H. (1949), *Applied Hydrology* (McGraw-Hill, New York), 689.

Lowry, W. P. (1967), *Weather and Life: An introduction to bioclimatology* (Academic Press, New York), 305.

Maunder, W. J. (1970), *The Value of the Weather* (Methuen, London), 388.

McLoughlin, J. B. (1969), *Urban and Regional Planning: A systems approach* (Faber and Faber, London), 331.

Mesarović, M. D. (1960), *The Control of Multivariable Systems* (Wiley, New York), 112.

More, R. J. (1967), Hydrological models and geography; In Chorley, R. J. and Haggett, P. (Eds.), *Models in Geography* (Methuen, London), 145–185.

Murphy, R. E. (1965), *Adaptive Processes in Economic Systems* (Academic Press, New York), 209.

Odum, E. P. (1959), *Fundamentals of Ecology*; 2nd End. (W. B. Saunders Co., Philadelphia), 546.

O'Riordan, T. and More, R. J. (1969), Choice in water use; In Chorley, R. J. (Ed.), *Water, Earth and Man* (Methuen, London), 547–573.

Rooney, J. F. (1967), The urban snow hazard in the United States; *Geographical Review*, **57**, 538–559.

Saarinen, T. F. (1966), Perception of the drought hazard on the Great Plains; *University of Chicago, Department of Geography, Research Paper*, **106**, 183.

Sewell, W. R. D., (Ed.) (1966), Human dimensions of weather modification; *University of Chicago, Department of Geography, Research Paper*, **105**, 423.

Stevens, J. C. (1938), The effect of silt removal and flow regulation on the regimen of the Rio Grande and Colorado Rivers; *Transactions of the American Geophysical Union, 19th Annual Meeting*, 653–659.

Strahler, A. N. (1958), Dimensional analysis applied to fluvially eroded landforms; *Bulletin of the Geological Society of America*, **69**, 279–300.

Terjung, W. H. (1970), Urban energy balance climatology; *Geographical Review*, **60**, 31–53.

Thomas, W. T. (Ed.) (1956), *Man's Role in Changing the Face of the Earth* (Chicago Univ. Press), 1193.

Van Dyne, G. M. (Ed.) (1969), *The Ecosystem Concept in Natural Resource Management* (Academic Press, New York), 383.

Walling, D. E. and Gregory, K. J. (1970), The measurement of the effects of building construction on drainage basin dynamics; *Journal of Hydrology*, **11**, 129–144.

Watt, K. E. F. (1964), Computers and the evaluation of resource management strategies; *American Scientist*, **52**, 408–418.

Watt, K. E. F. (Ed.), (1966), *Systems Analysis in Ecology* (Academic Press, New York), 276.

Wolpert, J. (1964), The decision process in spatial context; *Annals of the Association of American Geographers*, **54**, 537–558.

Glossary

Included in the following glossary are the terms which are, perhaps, those less familiar to physical geographers in that they are, in the main, derived from systems theory, statistical analysis, cybernetics, etc. The most useful sources for this glossary are listed below:

1 Cherry, C. (1966), *On Human Communication;* 2nd. Edn. (M.I.T. Press, Cambridge, Mass.), 337.
2 Clarke, D. L. (1968), *Analytical Archaeology* (Methuen, London), 684.
3 Cole, J. P. and King, C. A. M. (1968), *Quantitative Geography* (Wiley, London), 692.
4 Kendall, M. G. and Buckland, W. R. (1957), *A Dictionary of Statistical Terms* (Oliver and Boyd, Edinburgh), 493.
5 Miller, J. G. (1965), Living systems; *Behavioral Science*, **10**, 193–237 and 337–411.
6 Nagel, E. (1961), *The Structure of Science* (Routledge and Kegan Paul, London), 618.
7 Sippl, C. J. (1967), *Computer Dictionary and Handbook* (W. Foulsham and Co. Ltd., Slough, England), 766.
8 Young, O. R. (1964), A survey of general systems theory; *General Systems Yearbook*, **9**, 61–80.

ADDIVITY The condition when the variables under statistical investigation have no mutual effect or interaction upon each other. [3]*

ADJUSTMENT PROCESS CONTROL The operation of dominantly managerial attributes to produce the adoption of an adjustment decision model in the face of natural hazards (Kates, 1970) (See figure 8.6).

ALLOMETRIC GROWTH The growth of an organic system in dynamic equilibrium such that the ratios between each part of the system and the whole (and consequently between the parts) remain constant.

ASHBY'S LAW OF REQUISITE VARIETY For intelligence to gain control over a system it must be able to take at least as many distinct actions as the system can exhibit (Ashby, 1964).

BENEFIT/COST RATIO The comparison of the economic results achieved with the costs of a project or operation, expressed as a ratio. (In Britain expressed as a cost/benefit ratio.)

BIT One digit of a scale-of-two (i.e. binary) notation. Also a unit of measurement of quantities of information in communication theory. [1]

BY-PASSING The passage of mass, energy or information around some system component or store.

CANONICAL STRUCTURE The representation of an input/output process in terms of the simplest components which perform standard functions (Amorocho, 1965).

CAUSE Some process (i.e. a system input) which produces an effect.
Cumulative causation. An accumulation of causes.

* The numbers refer to the references at the beginning of the Glossary; other references are given to works listed at the end of the chapters.

346

CLUSTER ANALYSIS A technique for grouping variables into subsets that are linearly related in one fashion or another (Krumbein and Graybill, 1964).

COMPARATOR A device for comparing two signals and indicating whether they are equal or unequal. [7]

CONSCIOUSNESS The generation of new information *within* a system.

CORRELATION The average relationship between two or more variables, each of which is a series of measures of a quantitative characteristic. More loosely, the interdependence between variables such that when one changes so does the other, in a manner similar to that denoted by a mathematical function but not as explicitly defined. [2]

Autocorrelation. Correlation between members of a series of observations or samples ordered in space or time, which is due to a dependence between them. [4]

Multiple correlation. The average relationship or interdependence between more than two variables.

Coefficient of correlation (r). An index or measure of interdependence between two variables. It is usually a pure number which varies between $+1$ and -1.

Coefficient of determination (r^2). The square of the coefficient of correlation, expressing the proportion of variation in the dependent variable 'explained' by the association with the independent variable.

Coefficient of multiple correlation (R). An index or measure akin to r, but giving a precise value to the relationship involving more than one independent variable. The product-moment correlation between the actual values of the dependent variable in multiple regression and the values given by the regression equation. [4]

Coefficient of multiple determination (R^2). The square of the coefficient of multiple correlation, expressing the proportion of variation in the dependent variable 'explained' by the joint association with the independent variables.

CYBERNETICS The study of guidance and control problems in either biological or mechanistic systems. The science of control and communication in the machine and the animal. [7]

DAMPING The process of stifling or diminishing some effect. Sometimes used to describe the process of negative feedback or self-regulation.

DEGREE OF ADVANCEMENT The amount of change accomplished by a chemical reaction towards a new equilibrium state.

DEGREES OF FREEDOM The maximum number of variates that can be freely assigned before all the rest are completely determined. [3] Equal to the number of quantities which are unknown, minus the number of independent equations linking these unknowns (Blalock, 1964).

DEVIATION AMPLIFYING The 'snowballing' effect achieved by the operation of positive feedback.

DISPERSION The spread of a set of observations or objects (usually) about some central point or place. [3] The degree of scatter exhibited by observations about some central value. [4]

DISTRIBUTION The shape or apportionment of a population or sample of variates.

Normal distribution. Variates follow a normal distribution when they are grouped symmetrically about the arithmetic mean in such a way that 68.26% of them lie within $+$ or $-$ one standard deviation of the mean, 95.44% within $+$ or $-$ two standard deviations, and 99.73% within $+$ or $-$ three standard deviations.

Skewed distribution. One which departs from symmetry (Croxton and Cowden, 1948).

Left-skewed distribution. A distribution which is asymmetric, possessing a 'tail' of values to the left of the mode. Very infrequently encountered.

Right-skewed distribution. A distribution which is asymmetric in the opposite sense to left-skewed. One in which the median lies to the left of the mean, and the mode lies to the left of the median (i.e. there is a 'tail' of values to the right of the mean).

Logarithmic normal distribution. A right-skewed distribution which can be converted to the symmetrical normal form by plotting the logarithms of the values assumed by the variates.

Arrival distribution. A distribution of the probabilities that a given inter-arrival time in a queue of arrivals is greater than a given time (Gordon, 1969).

Inter-arrival time. The interval between successive arrivals in a queue, where T_a is the mean inter-arrival time (Gordon, 1969).

Erlang distributions. A family of arrival distributions representative of certain types of telephone traffic (Gordon, 1969).

EFFECT The result of a cause or a system input. In statistical analysis of variance it refers to the influence exerted over the average values assumed by the variable by each separate controlling factor or treatment.

ENDOGENETIC PROCESS One acting internally within the earth. [3]

ENTROPY In statistical thermodynamics, the expected logarithmic probability of the states of the thermodynamic system. The term is used, by analogy, in communication theory, to refer to the information rate of a source of messages. Used very broadly, it is a measure of the randomness of a system organization [1] —the probability of encountering given states, events or energy levels throughout the system.

Maximum entropy. A condition wherein there is an equal probability of encountering given states, events or energy levels throughout the system.

Low entropy. A condition of a system characterized by differentiation, organization and hierarchical structuring.

Negative entropy (*negentropy*). The tendency towards increasing system order. Free energy. Used in cybernetics as a synonym for 'information'.

Positive entropy. The tendency towards maximum entropy or the most probable distribution (i.e. towards maximum disorder or system sameness).

EQUIFINALITY The idea that the final state may be achieved from different initial conditions and in different ways (L. von Bertalanffy).

EQUILIBRIUM A condition in which some kind of balance is maintained.

Dynamic equilibrium. A circumstance in which fluctuations are balanced about a constantly-changing system condition which has a trajectory of unrepeated 'average' states through time.

Dynamic metastable equilibrium. A condition of dynamic equilibrium where thresholds allow occasionally large fluctuations to initiate new regimes of dynamic equilibrium.

Macroscopic equilibrium. The condition obtaining when, over a reasonable length of time, none of the observable macroscopic system properties changes appreciably.

Metastable equilibrium. The condition in which stable equilibrium obtains only in the absence of a suitable catalyst, trigger or minimum force which carries the system state over some threshold into a new equilibrium regime.

Poised equilibrium. The condition occurring when opposing forces are balanced.

Quasi-equilibrium. A state of near equilibrium. [3]

Stable equilibrium. A condition evidenced by the tendency for a system to move back to a previous equilibrium condition (i.e. to 'recover') after being disturbed by limited external forces.

Static equilibrium. The circumstance in which a balance of tendencies brings about a static condition of certain system properties, both absolutely and relatively.

Stationary equilibrium. The condition occurring when opposing velocities are balanced.

Steady state equilibrium. The condition of an open system wherein properties are invariant when considered with reference to a given time scale, but within which its instantaneous condition may oscillate due to the presence of interacting variables.

Thermodynamic equilibrium. The tendency towards a condition of maximum entropy in an isolated system.

Unstable equilibrium. A condition wherein a small displacement leads to a greater displacement, usually terminated by the achievement of a new stable equilibrium.

ERGODIC HYPOTHESIS The hypothesis that, under certain circumstances, space and time can be considered as interchangeable.

EXOGENETIC PROCESS One acting externally upon the earth. [3]

EXPLANATION The formulation in general terms of the conditions under which events of various sorts occur. [6] Used in correlation analysis to denote the coefficient of determination or multiple determination.

EXTERNALITY The link by which a component variable of a system is connected to an external variable.

FACTORIAL EXPERIMENT An experiment to determine the effects upon a variate of two or more factors (i.e. controls) when these are compared in all possible combinations of their several different levels. Used in conjunction with two- or n-way analysis of variance (Leclerg, Leonard and Clark, 1966).

FEEDBACK The feeding back of part of the output of a system as input for another phase of operation, especially for self-correcting or control purposes. [7]

Feedback loop. The path by which the feedback process is accomplished.

Deterministic feedback. Feedback process in which chance does not operate.

Negative feedback. Work done by the feedback mechanism which opposes the main driving force.

Positive feedback. Work done by the feedback mechanism which reinforces the main driving force, causing change to 'snowball' and the system to exhibit 'on-going' qualities.

FILTER A contrivance which allows the passage of certain magnitudes of mass, energy or information, while trapping or cutting off others.

Low pass filter. A means of treating a series in order to retain oscillations of long wavelength, while filtering out the short ones.

Space filtering. The process of treating spatial data so as to separate larger-scale regional trends from local variance (or 'noise).

Time filtering. The process of treating temporal data so as to separate longer-term trends from short-term fluctuations.

FOURIER SERIES A mathematical series consisting of terms containing sines and cosines used to represent certain complicated curves by means of an aggradation of simple wave forms described in sine and cosine terms (Harbaugh and Merriam, 1968).

Double Fourier series. An expansion of simple Fourier series so that the dependent variable becomes a function of *two* independent variables (which may be spatial co-ordinates), thus representing a complex surface instead of a curve (Harbaugh and Merriam, 1968).

FUNCTION A mathematical function is a quantity or dependent variable which takes on a definite value when a specified value is assigned to another quantity, quantities or independent variables. The mathematical expression of the dependent variable in terms of the independent variables so as to identify the relationship between them. [2]

Decision function. A mathematical function for making operational the information input by a decision maker into a feedback control system (see figure 8.3).

Exponential function. A mathematical function in which the dependent variable increases or decreases in geometrical progression, while the independent variable increases in arithmetic progression. This produces a relationship in which the rate of change of the dependent variable is always proportional to the corresponding value of the independent variable. Exponential functions are either positive or negative depending on whether the relationships are direct or inverse.

Objective function. A mathematical formulation of the present value of net benefits of a project.

Smoothing function. A mathematical expression employed to remove minor fluctuations from a curve.

Normal curve smoothing function. A smoothing function based on a normal distribution around the principal weight.

Negative exponential smoothing function. A smoothing function based on a negative exponential decay on either side of the principal weight.

GAME THEORY A set of mathematical models that deal with competitive problems, being widely used in military, industrial and bidding strategies. A method for the study of decision-making in conflict situations, where full control of the factors that influence the outcome is not possible.

GESTALT The concept that the significance of the whole is greater than that of the sum of its parts.

GOAL SEEKING The direction of a system operation towards some specified objective, such as a state, output level, etc.

HARMONIC ANALYSIS The analysis of a series of values into constituent periodic terms. [4] (See also Fourier Analysis.) The process of splitting a complex curve into its harmonic or periodic constituents, which differ in period and amplitude. [3]

HIERARCHY A set of items classified by rank or order. [3]

HISTOGRAM A diagram based on two axes perpendicular to each other with the horizontal axis divided into classes, and a column raised above each and measuring frequency proportionately on the vertical axis. [3]

HOMEOSTASIS The ability of a system to compensate for fluctuations in any part of the environment by means of negative feedback mechanisms. The tendency for a system to maintain a certain constancy of operation in the face of external fluctuations. (See also Self Regulation.)

Dynamic homeostasis. A self regulation the character of which changes with time.

INFORMATION The communication of constraint in variety between coupled systems or entities [2]. Knowledge or data, whose meaning is assigned by known conventions in its representation, not previously known to its receiver; and which is accurate, timely, unexpected and relevant to the subject under consideration [7].

INPUT External information, data, mass or energy transferred to storage (however temporary—i.e. throughput) within a system. [3]

INTERACTION System interaction: the exchange of variety between linked systems. *Statistical interaction*: in analysis of variance a significant interaction between two (or more) controlling factors occurs when the response of the dependent variable to variation in one factor is related in a non-parallel manner to variations in the other control(s), which interact.

ISOTROPIC Exhibiting equal properties or actions in all directions. When the properties or actions are unequal the medium is termed *anisotropic*.

ITERATION A process of successive approximations used in the numerical solution of certain complex equations.

KINEMATIC GRAPH A diagram showing transition probabilities associated with a Markov process.

LAG Retardation in a movement or transformation of any kind. The delay between two successive events. [7]

LEARNING Process of self-modification in response to (usually repeated) external stimuli (Ory). The retention of adaptive response patterns (Buckley, 1967).

LE CHATELIER'S PRINCIPLE If a change of stress is brought to bear on a system in equilibrium, producing a change in any of its components, a reaction occurs so as to displace the system state in a direction which tends to absorb the effect of the change, such that if this reaction occurred alone it would produce a change in an opposite sense to that of the original stress change.

LIEBIG'S LAW OF THE MINIMUM Originally stated that the growth of a plant is dependent on the amount of that nutrient which is presented to it in minimum quantity. Since expanded to include the effect of other physical factors and time (Odum, 1959).

LINGUISTIC INSTRUCTION The transference of information about the environment and of correct responses to it (Buckley, 1967).

MANIFOLD (In mathematics) A connected, locally-compact surface in an N-dimensional space. Elementary examples of manifolds are a closed curve in two dimensions and a sphere in three dimensions.
A topological space equipped with a family of local co-ordinate systems that are related to each other by co-ordinate transformations belonging to a specified class.

MARKOV CHAIN A system with a sequence of possible states such that the conditional probability of transition from one state to another does not depend upon how the former state was arrived at.
Markov process. Originally used to describe any process generating a stochastic series, the *adjacent* terms of which are related by given transition probabilities. Now extended to include stochastic series having statistical influence extending over *any* finite-length sequences. [1]

MATRIX An ordered array of numbers on which certain operations may be performed (e.g. multiplication). [3]

MEASURE OF CENTRAL TENDENCY A statistic or parameter indicating the degree of clustering within the distribution yielded by a sample or population.
Mean (Arithmetic). The sum of all values, divided by the number of values [3].
Median. The value that is central in an ordered series of values, having an equal number of values above and below it. [3]
Mode. The most commonly occurring value in a group of data. The class of values occurring most frequently.

MISSING PLOT TECHNIQUES Procedures used with two-to-n-way analysis of variance to estimate the values of missing observations (Leclerg, Leonard and Clark, 1966).

MODEL A representation of an event, object, process or system that is used for prediction or control. By manipulating the model the effects of changing one or more aspects of the entity represented can be determined. Several types of model have been identified, particularly, on the one hand, those which describe, and, on the other, those that reveal causal relationships (i.e. 'explain'). Mathematical equations are the most common example of the latter type.

Capture model. A stream network model which develops by the application of chance processes to the capture of channels and the migration of divides (A. D. Howard).

Deterministic model. A model whose behaviour is uniquely predicted by mathematical functions.

Geometric scale model. A model physically constructed so as to preserve a constant linear scale ratio between the model and the real world feature.

Growth model. A model which develops by a logical progression under the operation of deterministic processes.

Kinematic scale model. A model physically constructed so as to preserve a constant kinematic scale ratio (i.e. that involving velocities and accelerations) between the model and the real world feature.

Random model. A model which exhibits, at least in part, the operation of random (i.e. chance) processes. This is often termed a *stochastic model.*

Biased random model. A model which exhibits, at least in part, the operation of random (i.e. chance) processes which have unequal probabilities attached to them.

Rheological model. Mechanical analogue for the deformational behaviour of different states of a material, as it is subject to strain, in passing from one equilibrium state to another following the application of a new stress.

Simulation model. A model which simulates some property or properties of a real-world system.

Systems-based computer model. A model simulating a real-world system which is designed for operation with the aid of an electronic computer.

MORPHOGENETIC The deviation-amplifying effect whereby the morphology of a system conditions its subsequent development.

MULTIPLE-PURPOSE PROJECT A development plan designed to accomplish simultaneously two or more economic or social aims.

NOISE (In telecommunication) Disturbances which do not represent any part of the messages from a specified source. [1]

(In statistics) Unexplainable variance.

Random noise. Unexplained variance exhibiting no discernible pattern in space, in time, or in terms of any other variable.

Spatial noise. Spatially-distributed residuals from regression surfaces (i.e. unexplained spatial variance).

Standard error noise. The unexplained variance arising from the assumed normal distribution of scatter of Y about $Y_c = a + bX$.

White noise. A random, featureless noise, the values of which are independent of both the X and Y variables.

OPTIMUM The best result that can be obtained. [3]

Optimizer. One who takes a decision under assumptions of complete economic rationality, having a single profit goal, omniscient powers of perception, reasoning and competition, and who is blessed with perfect predictive abilities— in short, Economic Man! (Wolpert, 1964).

Optimum efficiency. Operating in such a manner as to fulfill an assigned role most economically and effectively.

OUTPUT The mass, energy, information or changes of state produced by the passage of an input through a system.

OVERSHOOTING The exceeding of some pre-determined goal or equilibrium state.

PALIMPSEST A surface bearing superimposed inscriptions of differing date.

PARAMETER Mathematical term for a symbolic quantity that may be associated with some measurable quantity in the real world (e.g. length).

Used in statistics to denote an invariant numerical feature of a population (e.g. size, mean, standard deviation, etc.).

PERCEPTION Observations that depend on the nature of the observer.

Environmental perception. Observations of the environment that depend on the nature of the observer.

PERSISTENCE EFFECT The tendency for events or values of similar magnitude to be grouped together in time.

PHASE A distinct type, region or economy of system operation, commonly demarcated by thresholds.

Gibbs' Phase rule. The rule governing chemical equilibrium relationships.

Phase function. An equation in which the average state of a complex system is expressed with reference to the variables defining the phase space.

Phase space. A hyperspace in which the states of a system may be represented, the axes of which represent a specific set of independent attributes (variables). [1]

PLEXUS A closely-interwoven network of intercommunicating lines.

PRINCIPAL WEIGHT The central value of a smoothing function.

PROBABILITY (In mathematics) A precisely defined value that indicates the chance (or odds) of an event occurring. The ratio of the number of ways in which an event can happen with respect to the total number of ways in which it can either happen or not happen.

Arithmetic probability paper. Graph paper having the scale on the ordinate so adjusted that normal distributions, when plotted in a cumulative manner, appear as straight lines, the angle of inclination of which is inversely related to the standard deviation of the distribution.

Extremal probability paper. Graph paper designed by E. J. Gumbel used to transform a plot of the magnitudes of extreme events or values against their return periods into a linear form.

Logarithmic probability paper. Graph paper having the scale on the ordinate so adjusted that logarithmic normal distributions, when plotted in a cumulative manner, appear as straight lines.

Probability distribution. A distribution of values of a variate showing the probability of encountering each value.

PROGRAMMING The act of planning a sequence of events or operations.

Computer programming. The devising of a series of logical operations to be performed by a computer.

Dynamic programming. A type of mathematical programming developed by the Rand Corporation in 1952 that involves a multi-stage process of decision making, wherein each consecutive decision must take into account its effects on later decisions to arrive at an overall optimum result.

Linear programming. An optimizing procedure enabling one to find the maximum or minimum value of a combination of linearly related variables subject to a number of constraints on the values which they may take.

Non-linear programming. An optimizing procedure enabling one to find the maximum or minimum value of a combination of non-linearly-related variables subject to a number of constraints on the values which they may take.

QUEUEING THEORY Mathematical techniques developed to deal with a class of problems arising from costs associated with bottlenecks and idle capacity, and used to determine the optimum size and design for production or service facilities. Concerned with the design of a queue discipline which will process an assumed arrival rate (expressed either in deterministic or stochastic terms) with the minimum connection delay.

QUARTILE There are three variate values which separate the total frequency of a distribution into four equal parts. The central value is called the median and the other two the Upper and Lower Quartiles, respectively. [4]

RANDOM Implies the equal chance of any of a number of events occurring, or of a variate being chosen. Generally the use of the term 'random' implies that the process under consideration is in some sense probablistic. [4]

REACTION The response to, or result of, action. The response of a system to an input.

Reaction potential. In chemistry: The driving force or impetus of a chemical reaction; being the ratio of the derivative of free energy and that of the degree of advancement.

Reaction time. The time period separating the input or change of input and the *beginning* of the resulting change in a system.

RECOVERY The return of a system towards its original state on the cessation of a short-term input.

REGRESSION A statistical method for investigating the relationships between variables by expressing the approximate functional relationship between them in an algebraic equation, or its graphic equivalent.

Linear regression. A type of regression in which the simple model (or regression equation) is in the form $Y_c = a + bX + E$, and that for more complex situations is $Y_c = a + b_1X_1 + b_2X_2 + \ldots b_mX_m + E$ (King, 1969).

Non-linear regression. Any algebraic equation in which the exponent of the independent variable (or variables) is greater than unity (see, for example, polynomial regression).

Multiple regression. A regression involving more than one independent variable.

Polynomial regression. A regression of the form $Y_c = a + bX + cX^2 + dX^3 + \ldots$ The 'degree' of the curve is designated by the exponent of X — e.g. a first-degree polynomial is the linear regression $Y_c = a + bX$; the second-degree polynomial is a parabola $Y_c = a + bX + cX^2$.

REGULATOR A component which tends to stabilize the system internally.

Decision regulator. A 'valve' which is the means by which socio-economic decisions can be implemented within a control system.

Self regulation. The operation of the negative-feedback process within a system.

REINFORCEMENT The increase of effectiveness of one variable as a result of its operation in conjunction with another.

RELAXATION The passage of a system from one equilibrium state to another.

Relaxation path. The sequence of states whereby adjustment to a new equilibrium is achieved following a change of input.

Relaxation time. The time taken by a system reorganization to achieve a new equilibrium, following a change of input.

REPLICATION Repeated sampling of a class of phenomena. The inclusion of the same mathematical information in two or more variables in one correlation analysis.

RESIDUAL The difference between an observed and a computed value. [3]

RETURN PERIOD (RECURRENCE INTERVAL) The average length of time separating events (i.e. the frequency of occurrence) of similar magnitude.

REVERSAL OF POLARITY When a high point of a curve is transformed into a low point, or *vice versa.*

SAMPLE A part of a statistical population, or a subset from a set of units, which is provided by some process or other, usually by deliberate selection, with the object of investigating the properties of the parent population or set. [4]

SATISFICER One who is content with limited, non-optimum economic goals; operating at an economic level which, although satisfactory to himself, is below the maximum.

SCATTER PLOT Conventionally used to describe the plot of paired values of dependent and independent variables upon a graph. Frequently used as a prelude to correlation analysis.

SECOND LAW OF THERMODYNAMICS The thermodynamic law stating that in an isolated system the change in entropy is always positive.

SECONDARY RESPONSE A system change which is not immediately prominent, either due to its small magnitude or the existence of a pronounced operational time lag.

SENSITIVITY Capacity for responding readily to stimulation.

SERVOMECHANISM A device to monitor the operation of a system as it proceeds, and to make necessary adjustments to keep the operation under control. [7] An automatic control device, such as a thermostat or automatic pilot, which corrects performance to a desired standard by means of an error-sensing feedback.

SIGNAL An event or phenomenon that conveys information from one point to another. [7] The physical embodiment of a message, utterance, transmission or sign-event. [1]

Error signal. The signal produced by a comparator in a servomechanism to modify the input in order to stabilize the system.

SIMULATION 'Feigning or imitation'. A way of representing real systems in an abstract form for purposes of experimentation, so that there is a close relationship between the experimental situation and the real-life situation which remains unaffected. Simulation is commonly performed by means of mathematical or physically-constructed models.

Computer simulation. To imitate system processes and features with mathematical models employing an electronic computer to perform calculations and logical operations (Harbaugh and Merriam, 1968).

SINGULAR POINT A location within a complex system at which a comparatively small applied force can effect relatively large transformations, because of the character of the point and its position relative to the whole system (J. Clerk Maxwell).

SOCIAL DECISION MAKING The pooling of ideas into a single environmental model (Buckley, 1967).

STANDARD DEVIATION The most widely used measure of dispersion of a frequency distribution; being equal to the positive square root of the variance. [4]

24

STANDARD ERROR *Of the mean:* A measure comparable with the standard deviation, which estimates the average dispersion of the means of samples of a given size about the population mean.

Of estimates: Used in correlation analysis as a measure of the average departure of the observed values of the dependent variable from those calculated (Croxton and Cowden, 1948).

STATISTICAL MECHANICS The formulation of statistical laws regarding the behaviour of complex systems composed of assemblages of small particles.

STATISTICAL STABILITY An equilibrium state or tendency the existence of which can only be recognised statistically by sampling a large number of features or events.

STATISTICS OF EXTREME EVENTS Statistical theory applied to the high or low extreme values of large populations of independent variates or events.

STEADY STATE The balance between inputs and outputs of energy and/or mass of a system maintained in such a manner so as to maintain its level of integration (Ory). (See *Steady state equilibrium.*)

STOCHASTIC Implying the presence of a random variable. A stochastic process is one wherein the system incorporates an element of randomness, as opposed to a deterministic system. [4]

STORE (American: STORAGE) A device capable of retaining mass, energy or information.

STRAIN The response of a system to the force (stress) acting upon it. [3]

STRESS The force acting upon a system. [3]

SYSTEM Any structured set of objects and/or attributes, together with the relationships between them. A whole which is compounded of many parts—an ensemble of attributes. [1]

Adaptive decision system. A system composed of a complex or hierarchy of feedback control mechanisms.

Black box system. A system which is analysed and treated as a unit, without any consideration of its internal composition or structure, so that attention is directed solely to the character of the outputs which result from identified inputs.

Cascading system. A system made up of a chain of subsystems, each having both spatial magnitude and geographical location, which are dynamically linked by a cascade of mass or energy.

Closed system. A system which is closed to the import and export of mass, but not of energy (Kern and Weisbrod, 1967).

Control system. A process-response system in which key components ('valves') are controlled by some intelligence, causing the system to operate in some manner determined by the intelligence.

Correlation system. A system identified on the basis of the significant correlations linking its components.

Ecosystem. Any system composed of physical-chemical-biological processes within a space-time unit of any magnitude (Lindeman). An area that includes living organisms and non-living substances interacting to produce an exchange of materials between the living and non-living parts (Odum, 1959).

Feedback control system. A system containing one or more negative feedback loops and capable of exerting different orders (on a scale of 0–5) of self-control.

Geographical system (Geographical control system). A dominantly spatial control system together with its non-spatial socio-economic inputs (see figure 8.26).

Grey box system. A system of which only certain limited internal structural and operational features are studied in order to obtain a better understanding of input/output relationships.

Human ecosystem. A system composed of interlocking social systems and ecosystems.

Human use system. A managerial unit capable of some measure of independent action with reference to external processes.

Irregular-surface system. Used here to designate a spatial process-response system in which flows of mass and energy possess spatially-variable horizontal and vertical components (e.g. a land surface).

Isolated system. A system with boundaries closed to the import and export of both mass and energy (Kern and Weisbrod, 1967).

Line system. Used here to designate a process-response system in which flows of mass and energy may be considered as essentially linear (e.g. a stream channel).

Morphological system. A system identified on the basis of the existence of significant correlations between its morphological properties.

Natural event system. A system involving the magnitude, duration, frequency and temporal spacing of natural events.

Open system. A system with boundaries which allow the import and export of both mass and energy (Kern and Weisbrod, 1967).

Plane-surface system. Used here to designate a spatial process-response system in which flows of mass and energy possess only spatially-variable horizontal components.

Point system. Used here to designate a process-response system considered on such a scale so that the inputs of mass and energy may be treated as being essentially vertical, relating to one point in space (e.g. a single plant).

Process-response system. A system formed by the intersection of morphological and cascading systems, the links between them being commonly provided by morphological components which either coincide with, or are closely correlated with, storages or regulators embedded in the cascading system.

Self-maintaining system. Commonly applied to a living system, particularly to the lowest forms of life.

Sequential decision system. A system which is capable of responding to a sequence of control decisions.

Social system. A complex of interdependencies between parts, components and processes of social phenomena.

Socio-economic system. A complex of interdependencies between parts, components and processes of social and economic phenomena.

Spatial subsystem. A subsystem which is defined in spatial terms.

Structural system. Used in the present context to define the morphological components of a feedback control system.

Subsystem. An individual unit of which a system is composed.

Supersystem. A complex of component systems.

White box system. An input/output system which is studied in an attempt to obtain a complete understanding of its internal structure and operation.

SYSTEMS ANALYSIS A term often used to describe the activities of a new profession of applied scientists concerned with the study of phenomena in systems terms. Sometimes employed more specifically to describe a phase of systems studies, following upon 'systems synthesis', in which consequences relating to costs, performance, etc., are deduced from alternative systems.

TELEOLOGICAL Directed or goal seeking.

THOUGHT The symbolic rehearsal of potential behaviour against a learned model of the environment (Buckley, 1967).

THRESHOLD A condition marking the transition from one state or economy of operation to another.

THROUGHPUT Mass, energy or information which passes through a system.

TIME SERIES A series of successive observations of the same phenomenon over a period of time. [3]
Stationary time series. A sequence of temporal values relating to a given phenomenon exhibiting no long-term or overall trend.

TRAJECTORY A sequence of system states.
Time trajectory. A sequence of system states through time.

TREND A continuous consecutive movement in a series or sequence of states, whether they may be spatial, time or abstract series. [2]
Trend surface. A mathematical expression fitted to data distributed in space, by means of a least-sum-of-squares regression, to produce a three-dimensional surface.
Polynomial trend surface. A regression of the form:
$$Z = a + bU + cV + dU^2 + eUV + fV^2 + \ldots \text{etc.}$$
where U, V are spatial co-ordinates fitted by the method of least squares to a spatial distribution of values of Z to produce a surface in three dimensions.

VALVE A component of a cascading system (commonly a storage or decision variable) which can be varied within certain limits by the intervention of some intelligence.

VARIABLE In general, any quantity or value which varies, or a quantity which may take any one of a specified set of values.
Decision variable. Variables the values of which are specified at the start of an economic analysis and which must be satisfied by the subsequent analysis.
Dependent variable. That variable whose variation is thought or seen to be constrained by the values assumed by other (i.e. independent) variables. Commonly designated by 'Y' in regression analysis.
Independent variable. A variable whose variation is considered to constrain the values assumed by the dependent variable. Commonly designated by 'X' in regression analysis.

VARIANCE The variance of a population (σ^2) or sample ($\hat{\sigma}^2$) is a measure of dispersion within the population or sample, expressed by the mean-square deviation of the distribution (i.e. the arithmetic mean of the squares of the deviations from the distribution mean). [2]
Covariance analysis. A statistical technique for treating the homogeneity of data that involve two or more concurrent and correlated variables unaffected by the treatment (Leclerg, Leonard and Clark, 1966).
Explained variance. (r^2; R^2). The regression or between-sample proportion of the total variance in correlation or variance analysis. It is assigned to the influence of the independent variable(s) or controlling factor(s).
One-way analysis of variance. The simplest form of variance analysis, in which between-sample variation is ascribed to the effect of only one controlling factor, which may have two or more levels.
Total variance. Corresponds exactly to VARIANCE.

Unexplained variance ($1-r^2$; $1-R^2$). The residual or within-sample (error) proportion of the total variance in correlation or variance analysis. It is assigned to normally-distributed random variation within the sample(s).

Variance analysis. A group of statistical techniques based upon Snedecor's 'F' or variance ratio test, in which the total variance of two or more samples is apportioned into that arising from variation between sample means and that due to within-sample variation.

VARIATE One measure or estimate of a variable (e.g. X_i represents the i th variate of the variable X).

VARIATION The variety of values assumed by a variate.

Coefficient of variation (V_c). A simple measure of variability within a population or sample, calculated as one hundred times the standard deviation divided by the mean.

Cyclical variation (*cyclical movements*). That exhibiting a roughly oscillatory change.

Periodic variation (*periodic movements*). That exhibiting a regular sequence of repetitions.

Relative variability (V_r). A dimensionless measure of dispersion equal to

$$\frac{\Sigma(X - \bar{X})}{N} . \frac{100}{\bar{X}}$$

VECTOR A compound entity having a definite number of components. [2] A matrix with either a single row (row vector) or a single column (column vector). In geometry, a line directed and proportional in length to scalar magnitude. [3]

Decision vector. Values forming the link between the decision function and structural system of a feedback control system (see Fig. 8.3).

Environmental vector. Environmental input values (structural environmental vector) or values input by the decision maker (decision maker's environmental vector) into a feedback control system (see Fig. 8.3).

Historical information vector. Part of a feedback loop of a feedback control system containing a record of past environmental observations by the decision maker in the form of stochastic data providing probability distributions as guides to the decision function (see Fig. 8.3).

Structural state vector. Values forming part of the negative feedback loop in a feedback control system (see Fig. 8.3).

Index

Bold figures indicate chapter sub-divisions. Italic figures indicate diagrams.